THE RETINA

THE

RETINA

AN APPROACHABLE PART OF THE BRAIN

JOHN E. DOWLING

THE BELKNAP PRESS OF HARVARD UNIVERSITY PRESS

Cambridge, Massachusetts, and London, England 1987

This book is printed on acid-free paper, and its binding
materials have been chosen for strength and durability.

Library of Congress Cataloging-in-Publication Data

Dowling, John E.
  The retina: an approachable part of the brain.

  Bibliography: p.
  Includes index.
  1. Retina.   2. Vertebrates—Physiology.   I. Title.
[DNLM: 1. Retina—cytology.   2. Retina—physiology.
WW 270 D744r
QP479.D65   1987   596′.019153   87-152
ISBN 0-674-76680-6 (alk. paper)

Designed by Gwen Frankfeldt

FOR JUDITH

Preface

UPON reaching his fiftieth birthday, Bertrand Russell is reputed to have said that it was about time he decided what to do with his life. As I approached that milestone and decided to write this book, it was not with the notion of seeking new vistas for the rest of my career. Rather, after studying the retina for more than 25 years, it seemed appropriate to look back on what has been learned about retinal mechanisms and to see whether I could construct a coherent picture of how the retina functions. A substantial amount of information has been gleaned from this tiny part of the brain over the past quarter century, but it is clear there is much more to learn. I look forward happily to the prospect of spending another 25 years investigating this fascinating tissue.

I came to the retina as an undergraduate at Harvard while a student with George Wald, the discoverer of the role of vitamin A in vision. My earliest research was mainly biochemical in nature, concerned with vitamin A deficiency and photoreceptor function. From there I was introduced to electrophysiology and anatomy, and gradually my interests widened to the rest of the retina. I moved to the Johns Hopkins University in 1964, where our group at the Wilmer Institute initially concentrated on anatomical studies and the retina's wiring, then on physiology and the electrical activity of individual retinal neurons. In 1971 I returned to Harvard and began to focus on pharmacological aspects and retinal neurotransmitters. Most recently much of my work has again been biochemical as my associates and I attempt to learn how neuroactive substances act on retinal cells. The approach of my research has thus come full circle.

When I began to study the retina, virtually nothing was known about the retinal synapses or their organization, the electrical responses of most of the retinal cells, or how neuroactive substances act on retinal neurons. Understanding on all of these fronts has advanced spectacularly since then. In many ways, the substantial progress in retinal research over the past three decades reflects the rapid development of neuroscience during this period. Before 1950 neuroanatomy, neurophysiology, and neurochemistry were quite separate disciplines, located in different academic departments, with minimal communication between them. Then, in the 1950s, with the development of electron microscopy and intracellular recording, came the recognition that anatomical studies at the cellular and subcellular levels could shed considerable light on physiological mechanisms and vice versa. The signal event in this regard may have been the almost simultaneous discoveries in 1952 of synaptic vesicles in nerve terminals (revealed by electron microscopy) and of the miniature synaptic potentials in muscle fibers (determined by intracellular recording). The anatomical findings provided a plausible explanation for the quantal nature of neurotransmitter release from nerve terminals onto muscle fibers, whereas the physiological experiments indicated a role for the structures that had been newly visualized. These data suggested that during synaptic transmission, packets of neurotransmitter molecules stored in the synaptic vesicles are released from the presynaptic nerve terminals.

In the 1960s, neuroanatomists and neurophysiologists collaborated closely, the term *neurobiology* was coined, and new interdisciplinary departments of neurobiology were formed. Neurochemistry was also advancing rapidly. But links between the chemistry and the anatomy-physiology were not forged until the 1970s, when it was discovered that neuroactive substances released at synapses can activate enzyme systems and modify postsynaptic neurons through biochemical processes. Today in neuroscience laboratories, anatomical, physiological, and chemical experiments are going on side by side, often carried out by the same investigator.

In this book I give an overview of much of what is known about the functional organization of the vertebrate retina, emphasizing an interdisciplinary approach. The major topics are those that have most interested me and on which my laboratory group has worked. Most of the illustrations are from work done in my laboratory, and I do not attempt to be inclusive; rather I have chosen those studies and those examples that I think have yielded the most useful and general insights concerning retinal function. I also speculate on the significance of the observations described whenever possible and present schemes as to how the retina may be organized anatomically, physiologically, and pharmacologically. I have also purposely limited the length and scope of the book to make it more readable and have tried to explain things as simply as possible, so that non-retinal specialists, students, and others interested in brain and visual mechanisms will find it of interest. I hope that the book conveys some of the excitement and enjoyment that I have experienced in studying the retina over the past two and a half decades.

After an introductory chapter, I discuss the cellular and synaptic organization of the retina (Chapters 2 and 3) and then consider the physiological and pharmacological properties of the retinal neurons (Chapters 4 and 5). The next two chapters are on the electroretinogram and glial responses and on visual adaptation (Chapters 6 and 7). In the final chapter I describe those studies that have had special significance for our understanding of brain mechanisms. A short primer is appended to aid the general reader, and the nonneurobiologist might find reading the final chapter helpful before plunging into the details of Chapters 2 through 7.

Needless to say, most of the ideas and facts presented in the following pages reflect the contributions of the marvelous coworkers I have had over the past three decades. They now number nearly a hundred and include undergraduate students, graduate students, postdoctoral fellows, visiting research fellows, and colleagues. I hope that they have learned as much from me as I from them.

THE FIRST draft of this book was written while I was on sabbatical leave in Okazaki, Japan, at the National Institute for Basic Biology. Ken-Ichi Naka and his colleagues were most welcoming, and a special note of thanks goes to them and the Japanese Society for the Promotion of Science for making my stay so pleasant. A number of individuals have given generously of their time, reading, criticizing, and providing many insightful comments on the manuscript, including Brian Boycott, Paul Brown, Joseph Dowling, Jr., Berndt Ehinger, Stuart Mangel, Robert Miller, Orlando Patterson, Ido Perlman, Harris Ripps, Hiroko Sakai, Frank Werblin, and Samuel Wu. I also appreciate the permissions to use figures from many sources and for the gifts or loans of the transparencies used for the color plates. Part of Chapter 5 was used earlier for an article in *Trends in Neuroscience* (9:236–240, 1986). My long-time associates, Patricia Sheppard and Stephanie Levinson, were indispensable, preparing the figures, typing the manuscript, and helping with all the details of the project. Howard Boyer and Jodi Simpson of Harvard University Press expertly and enthusiastically edited the manuscript. Finally, the National Eye Institute and the National Institutes of Health have generously supported my research over these many years, and I am very grateful.

Contents

T H E R E T I N A

Calligraphy by Judith Dowling of the Japanese kanji for retina: *Mōmaku*. *Mō*, the upper character, means "net"; *maku*, the lower character, means "membrane."

Approaches to the Brain

OVER the past half-century a great deal has been discovered about individual nerve cells. We have learned how neurons carry information, with graded potentials or with action potentials; how neurons transmit information, at chemical synapses or at electrical synapses; and how nerve cells receive stimuli, at synapses or at specialized sensory sites. For some years neurobiologists have focused much attention on understanding intact neural systems. They have been trying to discover how interactions between neurons in a brain lead to meaningful patterns of neural activity and ultimately to perception, behavior, and even intelligence.

An immediate problem faced by an investigator embarking on such studies is the extraordinary complexity of the vertebrate brain and the enormous number of neurons involved. In the human, for example, somewhere between 10^{10} and 10^{12} neurons are in the brain, and a single neuron can make 10^3 to 10^4 synaptic connections. How, then, is it possible to break down the complexity so that we can gain some understanding of higher integrative brain function? Two obvious ways are to work on either simpler nervous systems or on simpler parts of more complex brains, and both approaches have been and are being successfully taken.

Invertebrates: Simpler Nervous Systems

For simpler nervous systems researchers have turned to invertebrates whose brains contain many fewer neurons. Invertebrates such as the marine snail, *Aplysia,* or the leech are commonly used for integrative neurobiological research and have only 10^5–10^6 neurons (Nicholls and Baylor, 1968; Kandel, 1976). Furthermore, in many invertebrates, including the leech and *Aplysia,* the nervous system is distributed along the animal in separate ganglia each of which may contain only 1,000 to 2,000 neurons. These ganglia control specific behaviors exhibited by the animals, and one can analyze the neural circuitry underlying the behaviors in considerable detail. In many invertebrates the neurons are large, a characteristic facilitating intracellular recording; and many of the invertebrate neurons can be consistently identified in a ganglion, a situation permitting the investigator to record from a certain cell in animal after animal.

Although most invertebrates do not have the behavioral repertoire of vertebrates, they do exhibit many neural and behavioral phenomena similar to those in higher order organisms. For this reason they can yield important insights into higher neural processes. A striking example is found in the neurobiological analysis of the gill withdrawal reflex in *Aplysia*, carried out by Eric Kandel and his coworkers (Kandel, 1976). The gill withdrawal reflex is a defensive action: the animal rapidly withdraws its gill beneath a protective mantle sheath when the fleshy siphon, attached to the mantle sheath, is stimulated. The behavior is mediated entirely within one (the abdominal) ganglion that has about 2,000 neurons. By recording from and stimulating between pairs of cells, Kandel and coworkers demonstrated that the basic reflex is mediated by approximately twenty-four sensory neurons that innervate the siphon; six motor neurons that receive input from the sensory neurons and innervate the gill musculature; and three interneurons that receive input from the sensory neurons and provide excitatory or inhibitory input to the motor neurons, to the other interneurons, or back to the sensory neurons.

A particularly interesting aspect of the *Aplysia's* reflex is that it rapidly and strongly habituates (decreases in magnitude) when the siphon is stimulated repetitively. Furthermore, the reflex can be sensitized (enhanced) by stimulating the animal elsewhere on the body surface, such as the head. (The sensitization signal impinges on the reflex circuitry by way of other interneurons in the abdominal ganglion that form synapses on the sensory neuron terminals.) Both habituation and sensitization can be induced to last for hours, for days, or even for weeks by extended training sessions in which stimuli are presented repetitively to the animal. In this regard, the gill withdrawal reflex exhibits elementary forms of learning and memory.

An understanding of the circuitry underlying the reflex enables one to investigate the loci and mechanisms underlying habituation and sensitization. Research has shown, for example, that alterations in transmitter release from the presynaptic sensory neuron terminals underlie both habituation and sensitization in the short term. With habituation, a decrease in transmitter release causes a smaller postsynaptic response in the motor neuron; with sensitization, an increase in transmitter release from the sensory terminal causes an enhanced motor neuron response. Subsequent research revealed that biochemical changes in the sensory neurons appear to mediate the alterations in transmitter release from the terminals and, further, that structural alterations, presumably induced by the biochemical changes, occur in the sensory neuron synapses during long-term habituation and sensitization (Kandel and Schwartz, 1982; Bailey and Chen, 1983). These discoveries have provided important clues as to how complex brain phenomena such as memory and learning can occur.

Invertebrate systems also can be used to gain insights into brain perceptual phenomena, as was shown by the classic study of Keffer Hartline and Floyd Ratliff on the *Limulus* (horseshoe crab) eye (Hartline and Ratliff, 1957). The lateral horseshoe crab eye is a compound eye consisting of about 1,200 simple eyes (ommatidia), each containing about fifteen receptor (retinular) cells and one output (eccentric) cell. The receptor cells are all electrically coupled to the eccentric cell; so the graded potentials generated when light is absorbed in the photoreceptor cells pass into the eccentric cell, where they generate action potentials that carry the visual signal to the rest of the animal's brain. An individual ommatidium, however, is not independent; indeed, branches from one eccentric cell axon extend laterally to impinge on adjacent eccentric cell axons and inhibit them. This phenomenon is called lateral inhibition.

The effects of lateral inhibition on the firing frequency of an eccentric cell are shown schematically in Figure 1.1a. In the idealized experiment illustrated, the action potentials fired in the eccentric cell of an ommatidium, marked x, are recorded with wire electrodes hooked around the cell's axon. When a light stimulus is confined to ommatidium x, the steady-state firing frequency of the cell is proportional to stimulus intensity; the firing frequency shown reflects stimulation with a moderate light intensity (top trace). Inhibition between eccentric cells is demonstrated by illuminating ommatidium y and ommatidium x simultaneously. Under these conditions the steady-state firing frequency of the eccentric cell of ommatidium x is significantly reduced (middle trace) because of lateral inhibition.

The strength of the inhibition between eccentric cells depends on the level of activity in the interacting cells and the distance between them. The stronger the illumination, the stronger the inhibition exerted; and near neighbors affect one another more strongly than do distant ones. Furthermore, the inhibition is reciprocal; each eccentric cell inhibits nearby eccentric cells, and in turn it may be inhib-

1.1 a Idealized drawings of a *Limulus* lateral eye (*left*) and eccentric cell recordings (*right*). Recordings of action potentials from eccentric cells are made by dissecting out single axons and resting them on electrodes connected to an amplifier and recording device. Adapted from Ratliff (1965).

b Firing frequency of a single eccentric cell in response to a sharp dim-to-bright edge that was moved in stepwise fashion across the ommatidial array of a *Limulus* eye. Retinal distance refers to the location of the edge and is measured in numbers of ommatidia between the eccentric cell being recorded and the position of the edge. Because of lateral inhibition, the differences in firing frequency of the cell were accentuated adjacent to the dim–bright contour relative to differences in firing frequency some distance away from the border (that is, approximately 10 ommatidia away). Data redrawn from Barlow and Quarles (1971).

c A series of intensity steps that illustrates the Mach band phenomenon. Although the reflected intensity of each step is constant from one border to the next, it appears as if the steps are lighter along the border with a darker step and darker along the border of a lighter step. The Mach band phenomenon also accentuates the intensity differences between the steps. This is shown by obscuring the border between two steps with a pencil or other thin object. The two steps now appear much more similar in intensity.

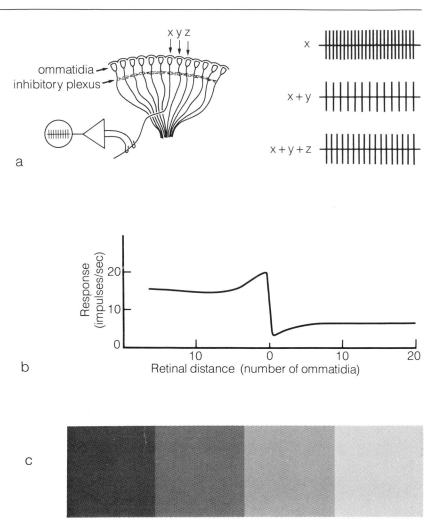

ited by them. An interesting consequence of the inhibitory interactions between eccentric cells is shown in Figure 1.1a when ommatidium *z* is illuminated at the same time as ommatidia *x* and *y*. The lower trace reveals that the discharge rate (firing frequency) of the eccentric cell of ommatidium *x* is now increased over that observed when only ommatidia *x* and *y* are illuminated. This happens because the eccentric cell of ommatidium *z* inhibits the eccentric cell of ommatidium *y*, and this inhibition causes a decrease of the inhibitory effect of *y* on *x*. This phenomenon is called disinhibition.

Hartline and Ratliff demonstrated that the simple lateral and reciprocal inhibition in the *Limulus* eye shapes the neural signals passing down the eccentric cell axons in such a way as to enhance contrast at an edge or border (Figure 1.1b). The phenomenon demonstrated in Figure 1.1b causes the discharge rates of eccentric cells to be higher along the lighter side of a stationary dim–bright border and lower along the darker side. This effect comes about because an eccentric cell in bright light but adjacent to a dark border is inhibited weakly by neighbors that are in dimmer light. Conversely, an eccentric cell along the dark side of such a transition is inhibited strongly by neighbors over the border in brighter light. Because of the interplay of excitatory and inhibitory influences, the differences in firing frequency between eccentric cells adjacent to a dim–bright border are significantly greater than they are for eccentric cells away from the contour. In other words, there is neurally fashioned border enhancement.

Enhancement of borders by the human visual system was recognized by Ernst Mach in the late nineteenth century and, consequently, is often called the Mach band phenomenon (Ratliff, 1965). Figure 1.1c shows a series of intensity steps that illustrates the Mach band phenomenon. Although each step is even in intensity from one side to another, it appears as if the steps are lighter along the margins bordering the darker steps, and darker along the margins bordering the lighter steps. We now believe that the Mach band phenomenon in ourselves is explained by a similar lateral and reciprocal inhibitory interaction occurring distally within the retina.

Vertebrates: The Visual System

The other approach to integrative brain mechanisms is to study well-defined and perhaps simpler parts of the vertebrate brain. The cerebellum, the olfactory bulb, and the visual system are brain regions that have been the subject of substantial research. The visual system has been particularly attractive, because it is easily stimulated with light and because it is naturally divided between eye and brain (Figure 1.2). Thus, the long optic nerve is especially accessible and enabled early investigators to learn something of how visual information is coded. In 1938 Hartline first recorded the activity of single cells in the visual system by dissecting out individual optic nerve axons. He showed that some fibers respond only during illumination, others at the termination of illumination, and still others at both the onset and offset of the light. Ragnar Granit (1947)

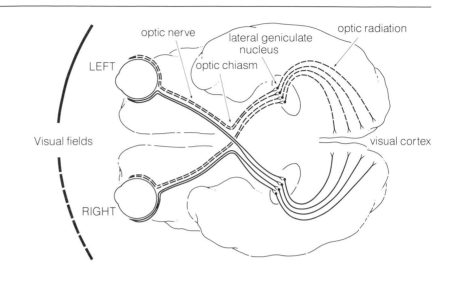

1.2 Diagram of the visual pathways in primates viewed from the underside of the brain. Visual information from the eyes passes via the optic nerve to the lateral geniculate nucleus, and from there, via the optic radiation, to the visual cortex. In primates the eyes face forward and the visual fields of each eye overlap. Information from the right visual field is received by the left half of each retina and vice versa. At the optic chiasm, the fibers from the right and left sides of the two retinas are sorted out so that information from the right visual field projects to the left side of the brain and information from the left visual field projects to the right side of the brain. Redrawn from Polyak (1941).

extended Hartline's work by introducing extracellular microelectrode recordings from the optic nerve and by demonstrating that similar kinds of responses could be recorded from ganglion cells in a wide range of animals. Horace Barlow and Stephen Kuffler's research in the early 1950s showed that single ganglion cells (the axons of which form the optic nerve) are transmitting significantly more information to higher visual centers (for example, lateral geniculate nucleus and visual cortex in mammals [see Figure 1.2]; tectum in cold-blooded vertebrates and birds) than the simple message that a light is turned on or off. Their experiments consequently set off an explosion of research on the visual systems of many kinds of vertebrates.

The fundamental discovery made at that time was that the response of a ganglion cell depended on the location of the illuminated point on the retina (Barlow, 1953; Kuffler, 1953). The same ganglion cell was excited when a light spot was positioned in one loci and inhibited when the illuminated spot was positioned elsewhere. (It is important to remember always that only the photoreceptor cells respond to light; whenever I state that a given cell in the retina or elsewhere is excited or inhibited by light, I mean that these effects are initiated by the photoreceptors.) Furthermore, excitatory and inhibitory zones interacted antagonistically: the cell responded weakly when both zones were stimulated simultaneously. Thus, the cell's response reflected an elementary spatial analysis carried out within the eye; in other words, the type of response elicited depended on the stimulus locus on the photoreceptor mosaic. These studies immediately suggested that further insights into brain processing could be gleaned by stimulating the photoreceptors with patterns of light and recording extracellularly from neurons along the visual pathway. The responses of ganglion cells of many species were systematically explored in this way, and David Hubel and Torsten Wiesel (1959, 1962, 1968) carried the analysis into higher brain centers, particularly into the cortical areas concerned with vision in cats and monkeys. This approach proved exceptionally rewarding: there appeared to be a hierarchy of cells in the visual cortex that, for optimal activation, required increasingly more specific stimuli to be presented to the eye. For example, to maximally activate optic nerve axons, spots of light projected on the retina are sufficient. To maximally activate cortical cells, however, oriented bars of light are required, which for some cells (called simple cells) may be stationary. Other neurons (complex cells) require oriented bars of light that are moving at right angles to the bar orientation, and some of these neurons respond only when the bar is moving in one direction (that is, these cells are direction sensitive). Still other more specialized complex cells respond maximally when a restricted and oriented bar of light, moving in a single direction, is presented to the retina. The conclusion from such findings is that along the visual pathway the visual image is broken down into components that are encoded in individual cells. Put another way, as information ascends to higher centers, there is increasing abstraction of the visual image in the neurons.

Virtually all of this research employed extracellular recordings, which provide information on the output of a neuron. With this technique the large, transient action potentials that carry information along axons are readily recorded, but the underlying, graded, synaptic potentials remain obscure. Thus, these studies yielded virtually no clues to how neuronal responses along the visual pathway are generated by interactions between cells. This information can be deduced only from intracellular recordings; but, for a variety of technical reasons, intracellular recordings from neurons are difficult to make from higher visual centers and from many regions of the vertebrate brain. Furthermore, the complexity of higher visual regions has so far precluded a satisfactory analysis of their cellular and synaptic organization—information vital for understanding the underlying circuitry for brain processes.

cornea   pupil   aqueous humor
iris
lens
vitreous humor
retina
optic nerve

1.3 Photomicrograph (*top*) and schematic drawing (*bottom*) of a primate eye.

The Approachable Retina

Unlike higher visual centers, the retina provides an excellent source of material for detailed anatomical, physiological, and pharmacological analyses of the neural mechanisms underlying elementary information processing by the vertebrate brain. The retina lines the back of the eye (Figure 1.3) and hence is very accessible. It has a well-defined, highly ordered anatomical organization and contains relatively few basic classes of cells. In many species, particularly cold-blooded ones, the cells are large, thus permitting intracellular recordings to be made routinely. A wide variety of ganglion cell responses have been recorded in many species, and some of these responses indicate a surprising complexity of information processing within the retina. Certain ganglion cells that have been recorded, particularly in nonmammalian vertebrates, have a response behavior comparable in complexity to some of the neurons in the visual cortex of mammals (Maturana et al., 1960; Maturana and Frenk, 1963).

With respect to its approachability for analysis, the retina can be regarded as the equivalent of an invertebrate ganglion. It is relatively isolated and is concerned with a specific and defined function. Its circuitry can and has been analyzed to a considerable degree, and the electrical responses of each of its cell classes are now reasonably well known. Yet the retina is a part of the vertebrate central nervous system, and thus mechanisms in the retina that dictate the response properties of neurons can serve as a model for neural mechanisms and interactions throughout the vertebrate brain.

The retina, like other regions of the vertebrate central nervous system, derives from the neural tube. Early in embryonic life the neural tube evaginates to form two optic vesicles in the head region of the embryo (Duke-Elder, 1963; Mann, 1964). Each optic vesicle subsequently invaginates to form an optic cup; and it is the neural epithelium on the inner wall of the optic cup that eventually becomes the retina (Figure 1.4.). Initially, both walls of the optic cup are one cell thick, but the cells of the inner wall divide to form a neuroepithelial layer many cells thick. These cells, called neuroblasts, differentiate into all of the retinal cells.

Photoreceptors form along the ventricular side of the neuroepithelial layer (Figure 1.4); thus, light must pass through the entire thickness of the retina to strike the photoreceptors. Cells on the outer wall of the optic cup differentiate into the pigment epithelium that lines the back of the eye. During development the photorecep-

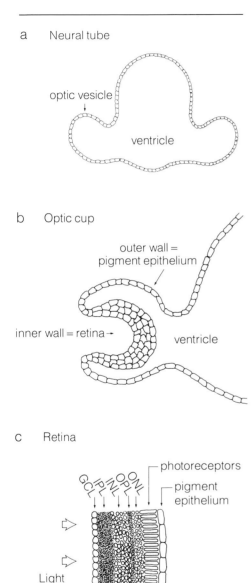

a   Neural tube

optic vesicle

ventricle

b   Optic cup

outer wall =
pigment epithelium

inner wall = retina

ventricle

c   Retina

photoreceptors

pigment
epithelium

Light

1.4 Schematic diagrams showing the optic vesicles, which develop from the neural tube (**a**); the invagination of an optic vesicle to form an optic cup (**b**); and the layers of the retina, which develop from the cells of the inner wall of the optic cup (**c**). Abbreviations defined in Figure 2.2.

tors and processes of the pigment epithelial cells interdigitate and thereby collapse the ventricular space between them. Tight cell adhesion almost never occurs between pigment epithelial cells and photoreceptors, however, and this fact has two consequences of note. First, in most species it is relatively easy to lift the retina completely intact from the back of the eye. The only points of firm attachment are around the periphery of the retina (ora serrata) and around the optic nerve. Once these attachment points are severed, the retina can readily be removed.

The other consequence is not so happy. Because of the lack of retinal attachment to the pigment epithelium, the retina can become detached from the back of the eye in a living animal by a blow to the head or eye. Vision is then lost in the detached part of the retina, and, unless measures are promptly taken, the entire retina often separates from the pigment epithelium, a condition resulting in complete blindness of that eye. Detached retinas are a serious clinical problem.

That the retina originates from the neural tube and is a true part of the brain is reflected in a number of its functional characteristics. For example, a blood–retinal barrier, apparently identical to the blood–brain barrier, restricts certain substances from gaining access to the retina from the vascular system. The retinal glial cells and their processes, like brain glial cells and their processes, fill most of the space between adjacent neurons, thus limiting extracellular space in the retina to narrow clefts a few hundred angstroms in diameter. How much extracellular space is in brain tissue is controversial, but arguments made in this regard for the brain apply also to the retina.

In many ways, the retina is an ideal part of the brain to study. As already noted, it can easily be removed intact from the back of the eye; and retinas isolated from cold-blooded vertebrates will usually remain viable for a few hours when kept moist and in an oxygen-rich environment. Furthermore, artificial media can be used to maintain retinas from both cold- and warm-blooded species for many hours. A similar and relatively simple anatomical organization is found in the retinas of all vertebrate species. The perikarya (cell bodies) of the neurons are segregated from the neuronal processes, forming discrete layers. The functional contacts (synapses) are made almost exclusively in two plexiform (process) layers, whereas the perikarya are distributed in three nuclear (cellular) layers (see Figures 2.1 and 2.2). In each plexiform layer the processes of a few classes of neurons interact, and thus it has been possible to identify by electron microscopy synaptic contacts made by the different classes of cells. The electrical activity of individual neurons can be recorded intracellularly; whereas with low-resistance electrodes placed on either side of the retina, field potentials can be recorded that reflect the activity of populations of cells and permit evaluation of overall responsiveness, sensitivity, and viability of the tissue. The retina's output is readily monitored by recording ganglion cell or optic nerve discharges with single-unit extracellular recording techniques that reflect the types of information processing that occurs within the synaptic layers. Also, it is easy to stimulate the retina naturally by focusing patterns of light on the receptors. Stimulus configuration, intensity, and wavelength can be readily manipulated, and the effects of these parameters have been investigated extensively.

Retinas from a wide variety of species have and are being studied. As might be expected, some retinas appear to be simpler than others and have proved to be particularly useful for the elucidation of basic retinal mechanisms. The mudpuppy retina, with its large cells and single types of cone and rod photoreceptors, has long been useful for studying the synaptic organization of the retina (see Figure 2.2). More recently the catfish retina, also with single types of cones and rods, has also yielded much useful information concerning retinal synaptic mechanisms.

Other retinas have features that make them especially appropriate for certain kinds of experiments or for studying special problems. The skate retina has only rod photoreceptors and has been used extensively to study mechanisms of visual adaptation. Regions of some retinas also have unique advantages. For example, in the central region of the primate retina, the small (midget) bipolar and ganglion cells in that part of the primate retina have limited axon terminal and dendritic fields, an arrangement making it possible to trace in the electron microscope the pre- and postsynaptic processes back to their cells of origin. Thus, the synaptic arrangements made between bipolar, amacrine, and ganglion cells were first recognized there.

Finally, there are retinas in which one cell type or another is particularly large or in which other features are especially advantageous. The extraordinarily large horizontal cell of the fish retina is a good example of the former, and many detailed studies of this cell type have been carried out. The interplexiform cell of the fish retina is an example of the latter. In teleosts, these cells contain and use as their transmitter dopamine, a transmitter substance about which we know a great deal. Moreover, we have several techniques for studying the properties of cells containing this agent. Indeed, much of what we know about the interplexiform cell has come from studies of the dopamine-containing cells in the teleost fish.

Retinal Cells and Information Processing

THE structure and organization of retinal cells has been studied intensively for over a century. Santiago Ramón y Cajal, the great Spanish neuroanatomist, early in his career recognized the value of the retina for the study of neuronal organization. In the introduction to his 1892 monograph *La rétine des vertébrés* he states: "Furthermore it is recognized that the retina is a genuine neural center, a sort of peripheral cerebral segment whose thinness, transparency and other qualities render it particularly favorable to histological analysis. In fact, although its cells and fibers are essentially similar to those of other centers, they are arranged in a more regular fashion, different types of cells being distributed in distinctly separate layers. Furthermore, the limited size of their dendritic fields, the orientation of the axon (always well defined and descendant) and the presence of layers arranged expressly for intercellular connections (outer and inner plexiform layers) are fortunate circumstances which help to clarify the morphology and interrelations of neurons" (Rodieck, 1973, p. 781).

Perhaps Cajal's greatest contribution to neuroanatomy was his realization of the enormous value of the silver-impregnation method discovered by Camillo Golgi in the 1870s. He improved upon Golgi's method and applied it to many parts of the nervous system, including the retina. The Golgi silver method usually stains entire nerve cells, from the finest dendritic branches to the axonal terminals. In any one section, however, only a small percentage of the cells are impregnated with the black silver precipitate, an outcome that permits the investigator to visualize individual nerve cells in exquisite detail.

Most of our present information on retinal cell types and on the extent and distribution of their processes has come from light microscopic analyses using the Golgi method. The many variants of this technique have been profitably employed by numerous investigators up to the present time, but by far the most influential studies were by Cajal in the late nineteenth century and by Stephen Polyak in the 1930s. Cajal described the retinal cells of many species, whereas Polyak focused his attention on primate retinas. Their books are still key references for any student of the retina (Ramón y Cajal, 1892; Polyak, 1941).

Cellular Organization

All vertebrate retinas are organized according to the same basic plan: two synaptic layers (outer and inner plexiform layers) are intercalated between three cellular layers (outer and inner nuclear layers and ganglion cell layer). Light, entering the eye, passes through the transparent retina and is captured by the visual pigment-containing outer segments of the most distal retinal cells, the photoreceptors (see Figure 2.1). Only the photoreceptors are sensitive to light. All visual responses in the retina (and elsewhere in the brain) are initiated by the photoreceptors. In addition to the photoreceptors, the retina has five other basic classes of retinal neurons: horizontal, bipolar, amacrine, interplexiform, and ganglion cells. There have been occasional reports of additional classes of neurons in the retina, but either these cells have been observed only in a single kind of retina (such as the stellate cell in the teleost retina [Ramón y Cajal, 1911] or the biplexiform cell in the primate retina [Mariani, 1982]) or they have been seen so infrequently that doubt exists as to their generality.

The perikarya of the photoreceptors are located in the outer nuclear layer, whereas the perikarya of four of the basic classes of retinal neurons are in the inner nuclear layer. Horizontal cells lie along the outer margin of the inner nuclear layer; the bipolar cell perikarya are predominantly located in the middle of the layer; and amacrine and interplexiform cell perikarya are arranged along the proximal border of the inner nuclear layer. The perikarya of the ganglion cells make up the most proximal layer, the ganglion cell layer.

Exceptions to this arrangement occur occasionally—when horizontal and bipolar cells are found in the outer nuclear layer, ganglion cells in the inner nuclear layer, and amacrine cells in the ganglion cell layer. Such cells are usually referred to as displaced cells, although it has now been shown conclusively that displaced amacrine cells found in the ganglion cell layer are a general and constant feature of many if not all retinas (Masland and Mills, 1979; Vaney et al., 1981).

The predominant type of glial cells in the vertebrate retina is called the Müller cell. These cells extend vertically through the retina, from the distal margin of the outer nuclear layer (or somewhat

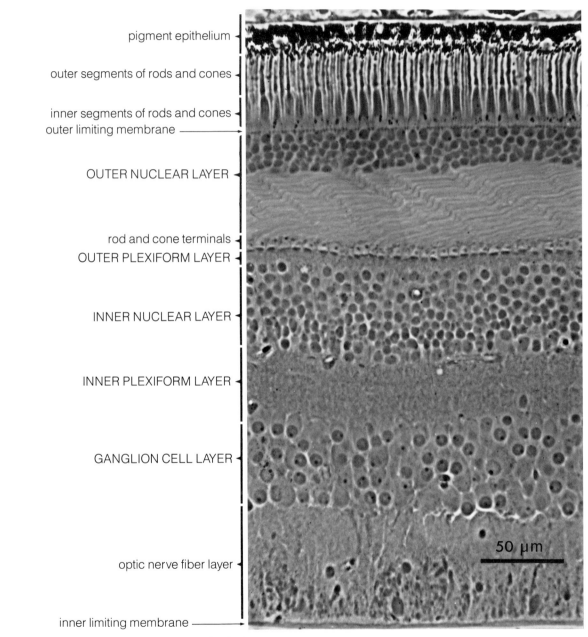

pigment epithelium

outer segments of rods and cones

inner segments of rods and cones
outer limiting membrane

OUTER NUCLEAR LAYER

rod and cone terminals
OUTER PLEXIFORM LAYER

INNER NUCLEAR LAYER

INNER PLEXIFORM LAYER

GANGLION CELL LAYER

50 µm

optic nerve fiber layer

inner limiting membrane

2.1 Vertical section through a human retina. The micrograph shows an area about 1.25 mm from the center of the fovea. In the foveal region of the retina the inner layers of the retina are pushed aside so that light can impinge more directly on the receptors. Thus, around the fovea and for some distance away (as shown here), the receptor terminals are displaced laterally from the rest of the photoreceptor cell (see Figure 2.4, which shows a foveal cone). Modified from Boycott and Dowling (1969), with permission of the Royal Society.

2.2 a Light micrograph of the mud-puppy retina showing the three nuclear layers, the two plexiform layers, and the prominent Müller (glial) cells (M). R, receptors, *olm,* outer limiting membrane; ONL, outer nuclear layer; OPL, outer plexiform layer; INL, inner nuclear layer; IPL, inner plexiform layer; GCL, ganglion cell layer; *ilm,* inner limiting membrane. From Dowling (1970), reprinted with permission of J. B. Lippincott Company.

b The major cell types found in the vertebrate retina, as viewed in a vertical section. This drawing is based on observations of cells found in the mudpuppy retina and impregnated by the Golgi method. R, receptors; H, horizontal cell; B, bipolar cells; A, amacrine cell; G, ganglion cell; I, interplexiform cell; M, Müller (glial) cell. Modified from Dowling (1970), with permission of J. B. Lippincott Company.

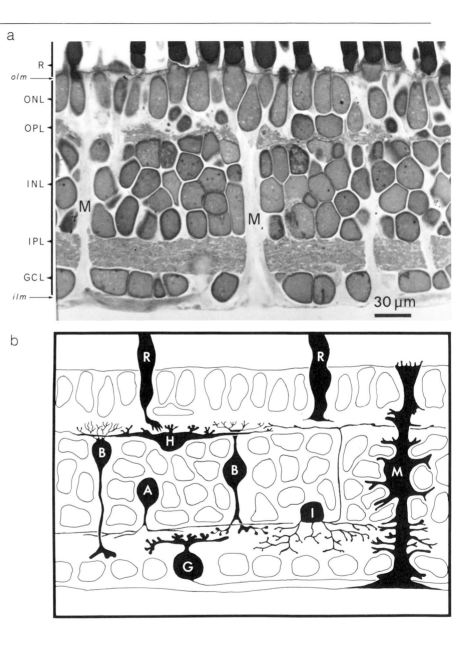

beyond) to the inner margin of the retina. The distal border of Müller cells is marked by the outer limiting membrane (Figure 2.2), which consists of junctional complexes made between the processes of different Müller cells or between the processes of Müller cells and photoreceptors (see Chapter 6). The proximal border of the cells is marked by the inner limiting membrane, which consists of the cell membrane of the Müller cell and a basement membrane. The nuclei of Müller cells are usually in the middle of the inner nuclear layer. Other types of glial cells have also been observed, particularly in the inner part of the retina; these include astrocytes (Ogden, 1978) and microglia (Boycott and Hopkins, 1981).

Figure 2.2a shows a light micrograph of the retina of the mudpuppy, an amphibian whose retinal cell perikarya are especially large. Note that the general organization of the mudpuppy retina is similar to that of the human retina shown in Figure 2.1; that is, both contain three cellular layers and two synaptic layers. There are, however, many fewer cells in the retina of the mudpuppy, even though the thicknesses of the two retinas are roughly comparable. Indeed, most vertebrate retinas are comparable in thickness. Retinas with larger cells tend to have fewer cells, whereas retinas with many cells tend to have smaller cells.

Figure 2.2b is a schematic diagram of the major cell types found in the vertebrate retina, showing their location in the nuclear layers and something of the shapes of the cells and the distribution of their processes within the plexiform layers. This drawing is based on preparations of the mudpuppy retina stained by the method of Golgi and shows that in the outer plexiform layer the processes of four classes of neurons interact. The receptor cell terminals (R) provide the input to this synaptic layer, whereas the bipolar cells (B) are the output neurons, carrying visual information from outer to inner plexiform layers. Horizontal cells (H) extend processes widely in the outer plexiform layer, but these processes are confined to this layer; they mediate lateral interactions within this first synaptic zone. The interplexiform cells (I) appear to be primarily centrifugal neurons, carrying information from the inner to the outer plexiform layers and spreading processes in both layers. The Müller (glial) cells (M) extend through the thickness of the retina from the outer to the inner limiting membrane; in the mudpuppy these cells are very prominent and are readily observed in conventional light microscopic sections of the retina (Figure 2.2a).

In the inner plexiform layer also, the processes of four neuronal classes interact. The bipolar cell terminals provide the input to the layer, and the ganglion cells (G) are the output neurons. Indeed, the ganglion cells are the output neurons for the entire retina; their axons run along the margin of the retina and collect at the optic disk to form the optic nerve, which carries all visual information from the eye to higher visual centers. Amacrine cells, like the horizontal cells in the outer plexiform layer, extend processes widely in the inner plexiform layer; their processes are confined to this layer and they mediate interactions within it. The interplexiform cells receive all of their input in the inner plexiform layer, and although they make some synapses in this layer, the majority of their output is in the outer plexiform layer (Dowling and Ehinger, 1978b). In nonmammalian species, centrifugal fibers that enter the eye via the optic nerve are often observed (Ramón y Cajal, 1892; Cowan and Powell, 1963). These fibers generally run through the inner plexiform layer and end along the border of the inner plexiform and inner nuclear layers on the amacrine and interplexiform cells and their processes.

All visual information passes across at least two synapses, one in the outer plexiform layer and another in the inner plexiform layer, before it leaves the eye. The situation is considerably more complicated than this, however, because in each plexiform layer considerable processing of visual information occurs. In most instances, therefore, visual signals go through many more than two synapses before they leave the eye. Thus the retina can be viewed as the equivalent of two brain nuclei: one representing the processing of visual information occurring in the outer plexiform layer, the other representing inner plexiform layer processing.

Classification of Retinal Neurons

Although there is general agreement that retinal neurons can be divided into six major classes, morphological studies, especially those employing the Golgi method, suggest that there are many cell types within each major class of cells. This is particularly true for amacrine and ganglion cells, for each of which investigators have reported as many as thirty morphologically distinct types in some species (see, for example, Kolb et al., 1981; Kolb, 1982). Physiological studies—particularly intracellular recordings—on the other hand, suggest that there are just a few basic response types for each class of retinal neurons, although some variations in responses are observed within classes. For example, two basic types of amacrine cell responses have been found in most retinas: sustained and transient responses (Chapter 4). Although it is usually easy to assign a particular response to one or the other of these categories, individual cells often show a mix of sustained and transient components. Thus, part of the anatomical heterogeneity may relate to various mixes of basic response types in different cells.

Insight concerning the large number of cell types recognized anatomically has come from pharmacological studies, which have shown that many neuroactive substances are found in the retina, particularly in the inner plexiform layer and in amacrine cells (Chapter 5). For example, more than twelve neuroactive substances have been identified with amacrine cells, and it has been possible to correlate a number of the morphological types of amacrine cells with specific neuroactive substances contained in these cells. These studies for the most part have employed immunohistochemical staining, which marks all the cells containing a particular substance, and have shown that many of the amacrine cell types are quite rare, each type consisting of only a few percent of the amacrine cell population or even less than 1 percent. Because Golgi silver-impregnation methods do not give reliable information concerning frequency of types of cells, the great majority of cells of a particular retinal cell class may consist of a few major types and a larger number of relatively infrequent types. There is good evidence for this in the cat retina, which has been claimed to contain twenty-three different types of ganglion cells (Kolb et al., 1981). On the other hand, it has been shown that one type (Boycott and Wässle's β cell, Kolb's G2 cell) makes up 55 percent of the total ganglion cell population in cats (Wässle et al., 1981a,b).

This is not to say that relatively rare types of cells may not play an important role in retinal function. A cell that makes up just 1 percent of a cell population can have very wide ranging processes and synapse on many other cells. Amacrine cells containing the peptide substance P are an example (Karten and Brecha, 1980). These cells make up only a tiny fraction of the total amacrine cell population in the retina, and their processes usually are confined to a very narrow stratum in the inner plexiform layer (Figure 2.3). But in that narrow layer the processes appear quite dense and could conceivably synapse on every process passing through the inner plexiform layer. One might speculate, however, that such a cell would not be involved directly in the processing of detailed visual information, but rather would be involved in modulating overall responsiveness or sensitivity of more discrete processing pathways. These notions will be discussed further in Chapters 5 and 8.

One reason that the morphologists often distinguish so many types among cells contributing processes to the inner plexiform layer is because of the layering or stratification seen in that layer in many species. This lamination was described first by Cajal (Ramón y Cajal, 1892), who noted that, particularly in the retinas of cold-blooded vertebrates and birds, at least five distinct strata can be distinguished. A number of cells appear to have their processes confined to one or a few strata (see Figure 2.3), and it is probably these highly stratified cells that underlie the lamination of the inner plexiform layer. Cells that are otherwise identical but extend their processes in one strata or another, or in a combination of strata, are often classified as separate types of cells. When additional criteria of branching pattern and size of dendritic spread are applied, investigators can distinguish a large number of cell types.

In mammals the stratification in the inner plexiform layer is not usually obvious, and thus its generality and significance has been questioned (Boycott and Dowling, 1969). It has been shown, however, that ganglion cells that respond to a spot of light when it is turned on (on-center cells) consistently have their processes in the lower part of the inner plexiform layer, whereas ganglion cells that respond to a spot of light when it is turned off (off-center cells) have their processes in the upper part of the layer (Famiglietti and Kolb, 1976; Nelson et al., 1978). Thus, there appears to be physiological significance for a division of the inner plexiform layer into at least two parts. Whether further subdivision of the inner plexiform layer has physiological significance remains to be determined.

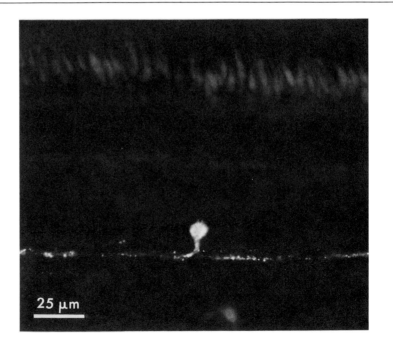

2.3 Fluorescence micrograph of a substance P-containing amacrine cell in the goldfish retina. The processes of this cell are confined for the most part to a narrow layer in the middle of the inner plexiform layer. The cell was stained immunohistochemically, that is, by an antibody that is specific to substance P and is linked to a fluorescent marker. Micrograph provided by C. L. Zucker.

25 μm

Some cautions with regard to the morphological classification of neurons should also be noted. Neuroanatomists sometimes distinguish two cells of the same class as different types solely on the basis of the dimensions of their dendritic fields (see Kolb, 1982), even though the cell processes of the two cells may extend along the same level of the inner plexiform layer, have similar branching patterns, and presumably have similar physiological responses. Many cells vary in their size, depending on where in the retina they are found; cells in the center of the retina, where visual acuity is high, are usually considerably smaller than cells in the periphery of the retina. Consequently, if cells only from the center and periphery of a retina are examined, they may appear strikingly different and be classified as separate types (see Boycott and Wässle, 1974). In addition, staining artifacts can cause misclassification. Occasionally cells impregnated by the Golgi method look quite different in the light microscope in terms of the appearance of their fine branches as a result of inconsistent deposition of the silver complex in the processes (Boycott and Dowling, 1969; Kolb, 1970).

In the following discussion of the classes of retinal cells, I shall describe only major cell types. This classification is based not only on morphology but also where possible on physiology. Clearly the situation is more complex than I suggest, particularly with regard to the amacrine and ganglion cells, but at the moment much of the anatomical complexity cannot be interpreted nor its significance understood. I shall return to this issue at the end of Chapter 3.

Receptor Cells

With one well-documented exception, all vertebrates so far examined appear to have at least two types of photoreceptors, which most often can be classified as rod cells and cone cells (Walls, 1942). Rods mediate dim-light vision, whereas cones function in bright light and are responsible for color vision. Classically rods and cones are differentiated on the basis of the shape of their outer segments, but this criterion is not always reliable. Thus, inner segment size and shape, terminal size and position, and perikaryal location in the outer nuclear layer are factors that may help in the identification process (see below). The exception noted earlier is the skate retina, which has only rod photoreceptors. But, interestingly, its receptors, although clearly rods morphologically, have physiological characteristics of both rods and cones; that is, these receptors function in both dim and bright light (Dowling and Ripps, 1970, 1972).

Usually there is a single type of rods in most retinas, but some animals, especially amphibians, are known to have two types of rods, called the red rods and the green rods on the basis of their color (Walls, 1942). Most species have multiple types of cones—often three, as in ourselves—which absorb maximally in the red, green, and blue regions of the spectrum (Marks et al., 1964; Brown and Wald, 1964). The difference in absorption is the result of the presence of different visual pigments in the outer segments of the cells (Plate 8; see Chapter 7). A few species (catfish and mudpuppy) appear to have only a single type of cones with a single visual pigment, and presumably these animals have little if any color vision. On the other hand, some species have a more complex situation. In birds and reptiles, colored oil droplets lie in the inner segment of many of the photoreceptor cells just adjacent to the outer segment (see Figure 2.4g,i), and these droplets serve as colored filters, modifying the spectral content of light reaching the outer segments (Walls, 1942). Thus, by combining different visual pigments with different colored oil droplets, birds and reptiles could theoretically have a large number of different spectral types of receptors. In fact, the combinations appear to be limited, so typically such retinas have at most five or six different spectral types of cone receptors (Ohtsuka, 1978, 1985; Kolb and Jones, 1982).

In fishes the spectrally different cones have different morphologies and thus can be distinguished anatomically (Scholes, 1975; Stell and Harosi, 1976). Double and twin cones are known in some species, in which two cones, either morphologically different (double cones) or similar (twin cones), are very closely apposed (Walls, 1942). In some species, rod and cone cells are difficult to distinguish, because the standard anatomical and physiological properties of the two types of photoreceptor cells are combined in one cell type. Some gecko photoreceptors are like this (Walls, 1942; Kleinschmidt and Dowling, 1975).

Figure 2.4 shows rod and cone cells from a number of animals and illustrates the diversity of photoreceptor morphology. All photoreceptors have an outer segment (which contains the visual pigment), an inner segment, a perikaryal region containing the cell nucleus, and a terminal. Typically rods have long outer segments and small, spherical terminals, but this is not always the case. Cones often have a shorter outer segment, a fatter inner segment, and a larger terminal than rods; but again, not all of these differences may exist in a particular species. From the receptor terminals of most cones and some rods, fine processes, or telodendria, are

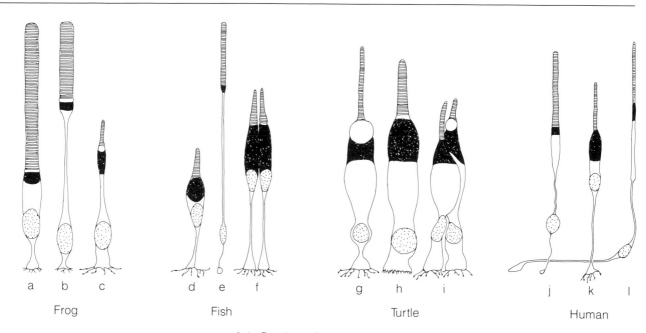

2.4 Drawings of rod and cone cells from a variety of animals. **a**, leopard frog red rod; **b**, leopard frog green rod; **c**, leopard frog cone; **d**, goldfish cone; **e**, goldfish rod; **f**, bluegill twin cone; **g**, snapping turtle cone; **h**, snapping turtle rod; **i**, western painted turtle double cone; **j**, human rod; **k**, human cone; **l**, human foveal cone. These drawings are modified from those of Walls (1942), with permission of Cranbrook Institute of Science.

observed to extend laterally to contact mainly other receptors. Short, finlike processes, observed by electron microscopy, extend outward from the inner segment of many receptors, both rods and cones; and in a few instances these processes have been shown to mediate interactions between the photoreceptors. Details of the fine structure of the photoreceptor terminals and their contacts will be given in Chapter 3, and a description of the outer segment region of the photoreceptor cell will be presented in Chapter 7.

Horizontal Cells

Most vertebrate retinas contains two basic types of horizontal cells: a cell with a short axon that typically runs 400 μm, or in some cases considerably further, before ending in a prominent terminal expansion, and an axonless cell. The axonless cell has not been observed in primate retinas; and it may be that other retinas are

2.5 Light micrograph of Golgi-stained horizontal cells from the cat, viewed by looking down on a flat mount of the retina. There are two types of horizontal cells in the cat retina: an axonless cell (**b**) and a cell with a short axon (**a**). From Fisher and Boycott (1974), reprinted with permission of the Royal Society.

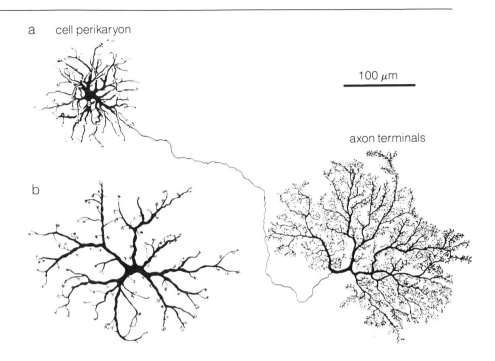

a   cell perikaryon

100 μm

axon terminals

b

lacking one of these horizontal cell types (Gallego, 1982). On the other hand, in some retinas there appear to be varieties of one or the other type of horizontal cells (Gallego, 1982).

Horizontal cells of the cat have been studied in particular detail. Figure 2.5 shows a drawing of both types from that animal. The cells are stained by the Golgi method and are observed by looking down on a flat mount of the whole retina (Boycott et al., 1978). Electron microscopy of Golgi-stained horizontal cells from the cat has shown that the processes of the axonless cell and the dendritic processes of the short-axon cell connect exclusively with cone terminals, whereas the axon terminal processes of the short-axon cell end exclusively in the rod terminals (Kolb, 1974). In other retinas such a clean segregation of cone and rod inputs into the horizontal cells has not been observed. The axon terminals in the turtle appear to receive input from both rods and cones (Leeper and Copenhagen, 1982). In the ground squirrel, on the other hand, input into axon terminals may come exclusively from cones (West, 1978).

Horizontal cells respond to illumination of the retina with sustained graded potentials; in situ they do not ordinarily generate action potentials. Thus, the role of the axon in the short-axon horizontal cell is somewhat puzzling, because responses generated either in the dendritic or axon-terminal processes should be severely attenuated before reaching the other end of the cell. If there is a segregation of rod and cone inputs into the cell, as in the cat, one might suppose that the axon provides for a segregation of rod and cone responses in different regions of the cell, and this is an appealing suggestion (Nelson et al., 1975). In those cases where both ends of the cell receive cone input, however, it is more difficult to understand the role of the axon. Furthermore, recent evidence suggests some mixing of rod and cone signals in the cat retina at the level of the photoreceptors, a mixing mediated by interreceptor contacts; so under some conditions one can record signals from both kinds of receptors in either the dendritic or the axonal end of the short-axon horizontal cell (Nelson, 1977). These observations also raise the question of the role of the axon in these cells.

The horizontal cells of the teleost fish retina are particularly large. They form several distinct layers along the distal margin of the inner nuclear layer, and their physiological characteristics have been studied extensively. Indeed, they were the first retinal cells to be recorded intracellularly. They are somewhat different in their organization from the horizontal cells described so far (Parthe, 1982), and so I shall give them special mention. In fishes, as in other vertebrates, there appear to be two basic horizontal cell types; however, the short-axon cells appear to connect exclusively with cones and the axonless cells exclusively with rods. Furthermore, in most fish retinas there are subtypes of short-axon horizontal cells that can be distinguished both morphologically and physiologically (Stell and Lightfoot, 1975). Figure 2.6 shows examples of the three subtypes of cone-related, short-axon horizontal cells found in the goldfish retina; these subtypes have been called H1, H2, and H3.

The axon terminals of the cells shown in Figure 2.6 do not have fine processes that connect with photoreceptors. Rather the axon terminals of these fish horizontal cells extend through the inner nuclear layer to end among the amacrine cells, or, occasionally, at the border of the inner nuclear and inner plexiform layers (Stell, 1975; Sakai and Naka, 1986). Some synaptic contacts are made by these terminals onto cell perikarya and processes in the inner nuclear layer and onto processes in or close to the inner plexiform layer (Sakai and Naka, 1986; Marshak and Dowling, 1987). Also a few synapses made onto the axon terminals by processes running in the inner nuclear layer have been observed. The role of the axon terminals in fish horizontal cells is unclear (but see Marshak and

2.6 Golgi-stained horizontal cells from the goldfish retina, viewed in flat mount. All three of these short-axon cells are exclusively cone-related. Note that, unlike the axon terminals of the short-axon horizontal cells in the cat (Figure 2.5), the axon terminals of these cells do not have fine processes that connect with receptor terminals. From Stell (1975), reprinted with permission of Alan R. Liss.

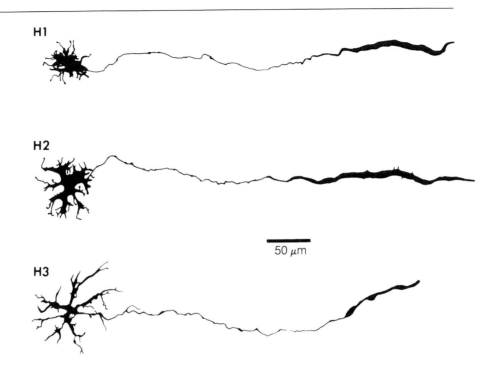

H1

H2

50 μm

H3

Dowling, 1987), although responses have been recorded from them; and there is some evidence that the axon terminals in the catfish can directly stimulate amacrine cells (Sakai and Naka, 1985).

Bipolar Cells

Bipolar cells have been analyzed most satisfactorily and completely in primate and cat retinas (Boycott and Dowling, 1969; Kolb, 1970; Boycott and Kolb, 1973; Mariani, 1981). There, three major types of bipolars have been distinguished: one type exclusively connected to rods and two types exclusively connected to cones (Figure 2.7). The two cone-related types of bipolar cells make different kinds of synaptic connections with the cones, their axonal terminals end in different parts of the inner plexiform layer, and they appear to be related to the generation of either on- or off-responses to light in the retina.

One type of the cone-related bipolar cells extends dendrites into invaginations of the cone receptor terminal, and the axon terminals of these cells end deep in the inner plexiform layer (on-response region). This type is called the invaginating bipolar cell. The other

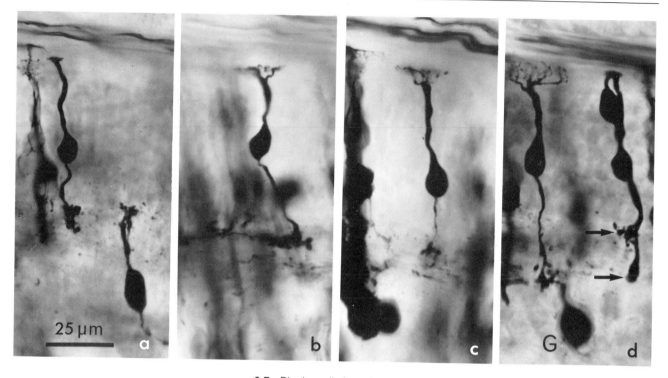

2.7   Bipolar cells from rhesus monkey retinas stained by the Golgi method.

a An invaginating cone (midget) bipolar cell (*left*) and a midget ganglion cell (*right*).

b, c Cone-related flat bipolar cells.

d A rod-related bipolar cell (*left*) and two cone-related midget bipolar cells (*right*).

Note that in **d** the dendrites of the two midget bipolar cells overlap, an arrangement suggesting that these two bipolar cells contact the same cone receptor terminal. Note also that the terminals of the two bipolar cells end on different levels of the inner plexiform layer (*arrows*), an indication that one is an invaginating midget bipolar cell and the other a flat midget bipolar cell (see text). G, an incompletely stained ganglion cell. From Boycott and Dowling (1969), reprinted with permission of the Royal Society.

cell type makes contacts along the flattened base of the cone receptor terminals, and the axon terminals end in the distal part of the inner plexiform layer (off-response region). This type is called the flat bipolar cell. After Golgi staining, these two types of bipolar cells in primates and cats can be distinguished in the light microscope on the basis of the appearance of their dendritic arbors and the position of their axonal terminals in the inner plexiform layer (Figure 2.7d). The rod-related bipolars extend their dendrites into invaginations of the rod terminals, and the axon terminals of these cells end deep in the inner plexiform layer. The dendritic arbor of the rod-related bipolars is considerably larger than those of the cone-related bipolars, and a single rod-related bipolar may contact as many as forty-five rod terminals. These cells, too, can easily be distinguished by light microscopy (Figure 2.7d).

In the primate retina some bipolar cells contact only a single cone terminal (Polyak, 1941). These cells, called midget bipolar cells, are found most commonly in the central (macular) region of the retina,* and they are thought to mediate the high acuity of the central cones in the primate. Midget bipolars have also been observed in retinas of birds and reptiles (Ramón y Cajal, 1911) and in the cone-dominated squirrel retina (West and Dowling, 1975). Both invaginating and flat midget bipolars are found in primate retinas and have been observed contacting the same cone terminals (Figure 2.7d).

In other retinas, especially in retinas of the cold-blooded vertebrates, the situation is not so clear. It has been shown that bipolar cells in general make two kinds of contacts with receptor cells, analogous to the invaginating and flat contacts made by primate and cat cone-related bipolar cells (see Chapter 3), and also that the bipolar terminals of a particular cell are usually found in one half of the inner plexiform layer or the other, but not in both. In most cold-blooded vertebrates, however, it is not possible to distinguish flat and invaginating types of bipolar cells on the basis of the appearance of their dendritic arbors in the light microscope (for reasons given in Chapter 3), and often the axon terminals of the bipolar cells are highly stratified, ending on one or several levels in the inner plexiform layer (Kolb, 1982).

Most cold-blooded vertebrate retinas contain large bipolar cells, long thought to be rod-related bipolar cells, and small bipolar cells, believed to be cone-related bipolar cells. In fishes it has been shown that many bipolar cells, particularly the larger ones, contact both

---

* Many primates and some birds have a high-acuity region of the retina that is characteristically specialized (Ramón y Cajal, 1911; Polyak, 1941). This region is called the fovea, and the inner layers of the retina are pushed aside from this area (typically 0.5–1 mm in diameter) so that light can impinge directly on the receptors. Only cones are present in this area in primates, and the primate foveal cones are the thinnest and the longest receptors in the retina (see Figure 2.4). No blood vessels are found in the fovea; and furthermore, there appear to be no blue-absorbing cones in the center of the fovea in primates. These specializations serve to improve visual resolution in this retinal area.

rods and cones (Stell, 1967; Scholes, 1975). It has also been shown that certain of these bipolar cells are likely to be color-coded, because they contact specific sets of cones (Ishida et al., 1980).

Amacrine Cells

The name amacrine cells was given by Cajal to cells that have no axon. He observed such cells not only in the retina but also in other parts of the brain. In the retina none of the amacrine cells appear to have axons or axonlike processes, with the possible exception of the so-called association amacrine observed by Cajal (Ramón y Cajal, 1892, 1911) in the bird retina; this amacrine cell type may be a short-axon amacrine cell similar to the short-axon horizontal cell. Usually all of the processes of a single amacrine cell look similar when viewed in Golgi-stained material by light microscopy (Plate 2). As already noted, amacrine cells are very diverse in their morphology, particularly in terms of the extent and distribution of their processes; and it is possible to describe a large number of amacrine cell types in most species.

Cajal (Ramón y Cajal, 1892) classified amacrines into two major types: diffuse amacrine cells (sometimes called fusiform amacrine cells) and stratified amacrine cells (sometimes called pyriform amacrine cells). The diffuse amacrine cells extend their processes throughout the thickness of the inner plexiform layer, whereas the stratified amacrine cells extend their processes on one or a few strata of the inner plexiform layer (Figure 2.8). This simple classi-

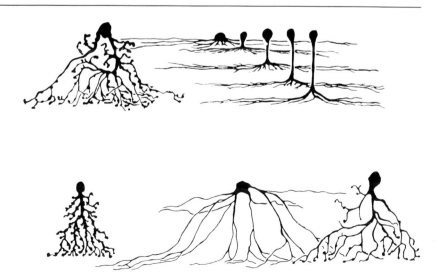

2.8   Drawings of mammalian amacrine cells, mainly from ox and dog retinas. *Top,* A diffuse amacrine cell and five stratified amacrine cells; *bottom,* three diffuse amacrine cells with different dendritic spreads. The cell on the bottom left is a narrow-field diffuse amacrine cell; the cell in the bottom middle, a wide-field diffuse amacrine cell; the cell on the bottom right is a diffuse amacrine cell with an intermediate-size field. From Boycott and Dowling (1969), who redrew these cells from Cajal; reprinted with permission of the Royal Society.

fication scheme can be expanded to include narrow- and wide-field diffuse amacrine cells (on the basis of the dendritic spread of the cells) and mono-, bi-, or multistratified amacrine cells (on the basis of whether the cell processes are confined to one, two, or several levels in the inner plexiform layer). And, of course, there are some amacrine cells that do not fit neatly into either subdivision; they have been given names such as stratified diffuse amacrine cells (Boycott and Dowling, 1969).

In general, the amacrine cells with the largest lateral dendritic spread in the vertebrate retina are the stratified cells, and the density of branching of the wide-spreading amacrine cells is less than that found for smaller-field cells. Furthermore, the retinas of cold-blooded vertebrates and birds usually have more kinds of multistratified amacrine cells than do mammals. It is also true, however, that many cold-blooded vertebrates and birds have a higher ratio of amacrine cells to other retinal neurons than do most mammals; the consequences of this will be discussed further in Chapter 3.

Interplexiform Cells

Interplexiform cells have been recognized as a distinct class of retinal neurons in the vertebrate retina for only about a decade. They were first clearly observed in the teleost retina in preparations processed by the Falk-Hillarp method (Plate 6), which causes those cells containing certain monoamines to fluoresce (Ehinger et al., 1969). Subsequently, they were seen in Golgi-stained material from the cat retina and were given the name interplexiform cells in recognition of the fact that they extend processes into both plexiform layers of the retina (Gallego, 1971). They have now been seen in both Old and New World primates, rabbits, toads, and skates (Boycott et al., 1975; Oyster and Takahashi, 1977; Kleinschmidt and Yazulla, 1984; Brunken et al., 1986).

These cells were recognized only recently because they are refractory to Golgi staining. Thus, Cajal only saw an occasional cell resembling the interplexiform cells, and subsequent authors using Golgi-stained material also failed to observe them. By searching specifically for them, however, it is possible to find these cells regularly. In our study of the cat and monkey retinas (Boycott et al., 1975) we saw over 100 interplexiform cells; although even in the best material the quality of interplexiform cell staining was usually poor. Why interplexiform cells should be impregnated poorly by Golgi method is not at all clear. However, this example does point out our enormous dependence on the Golgi method for identifying

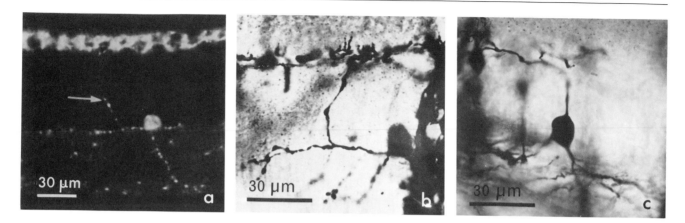

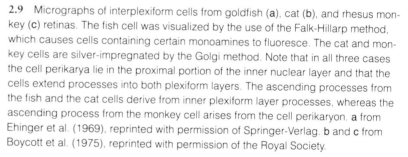

2.9 Micrographs of interplexiform cells from goldfish (**a**), cat (**b**), and rhesus monkey (**c**) retinas. The fish cell was visualized by the use of the Falk-Hillarp method, which causes cells containing certain monoamines to fluoresce. The cat and monkey cells are silver-impregnated by the Golgi method. Note that in all three cases the cell perikarya lie in the proximal portion of the inner nuclear layer and that the cells extend processes into both plexiform layers. The ascending processes from the fish and the cat cells derive from inner plexiform layer processes, whereas the ascending process from the monkey cell arises from the cell perikaryon. **a** from Ehinger et al. (1969), reprinted with permission of Springer-Verlag. **b** and **c** from Boycott et al. (1975), reprinted with permission of the Royal Society.

neuronal cell types in the brain. Thus, if a cell type is not stained well by this method, it may be overlooked.

Some authors have suggested that interplexiform cells should be considered not as a separate class of retinal neurons but as a type of amacrine cells (Witkovsky, 1980). The interplexiform cell perikarya do sit among the amacrine cell perikarya. But, unlike amacrine cells, they extend processes into both plexiform layers of the retina, and in some species the density of processes is greater in the outer plexiform layer than in the inner plexiform layer (Plate 6). Furthermore, the interplexiform cells appear to be mainly a centrifugal type of neuron, carrying information from inner to outer plexiform layers (see Chapter 3). For these reasons, it seems appropriate to classify these cells as a separate class of retinal neurons.

Processes of the interplexiform cells may ascend to the outer plexiform layer either from the cell perikaryon or from interplexiform cell processes in the inner plexiform layer. Figure 2.9 shows interplexiform cells from fish, cat, and monkey retinas. The ascending processes in the fish and cat retinas are from inner plexiform

2.10 Micrographs of Golgi-stained ganglion cells from the ground squirrel retina.

a, b Narrow-field stratified cells.

c A presumed midget ganglion cell.

d A wide-field stratified cell.

e A bistratified ganglion cell.

The scale bars all represent 30 μm. From West and Dowling (1972), reprinted with permission; copyright 1972 by the AAAS.

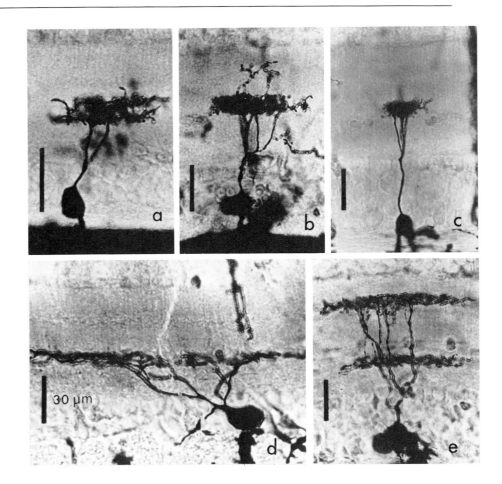

layer processes; the ascending process in the rhesus monkey retina is from the cell perikaryon. Not enough well-stained interplexiform cells have been observed yet to decide whether there are different types of these cells. On the basis of pharmacological evidence, Marc and Liu (1984) have proposed that there are two types of interplexiform cells in the goldfish retina: one type that contains dopamine and another that accumulates glycine.

Ganglion Cells

Ganglion cells, like amacrine cells, can be subdivided into two general types on the basis of the morphology of their dendritic trees (Figure 2.10): those with diffuse dendritic trees that spread throughout the inner plexiform layer and those with stratified dendritic trees that spread on one or a few levels of the inner plexiform layer (Ramón y Cajal, 1892; Boycott and Dowling, 1969). Most retinas seem to have more stratified ganglion cells than diffuse cells. Like amacrine cells, ganglion cells can be subdivided further into small- and large-field cells, mono-, bi-, and multistratified cells, and various combinations. Again, like amacrine cells, multistratified cells are more numerous in cold-blooded vertebrates and in birds than in mammals.

A particular type of ganglion cells found in the central region of the primate retina deserves special mention. The midget ganglion cell was first described by Polyak (1941), who pointed out that a midget ganglion cell may receive all of its input from a single midget bipolar cell. The dendritic spread of the midget ganglion cell is so limited that it may contact the terminals of just one midget bipolar cell (Figure 2.11); at most, a midget ganglion cell can receive input from only a few midget bipolar cells. Furthermore, midget ganglion cells come in two varieties: those whose dendrites ramify in the lower part of the inner plexiform layer and those whose dendrites

2.11   Golgi-stained midget ganglion cells from primate retinas.

a Two adjacent midget ganglion cells from the human retina. Note that the dendrites of the two cells end at two different levels within the inner plexiform layer (*arrows*). The arrowhead indicates the axon of the cell on the left, which is in better focus.

b The dendrites of a rhesus monkey midget ganglion cell in close association with the terminal of a midget bipolar cell.

c Another example of a rhesus monkey midget ganglion cell with its dendrites embracing a midget bipolar cell terminal. On the left is a terminal of another midget bipolar cell. From Boycott and Dowling (1969), reprinted with the permission of the Royal Society.

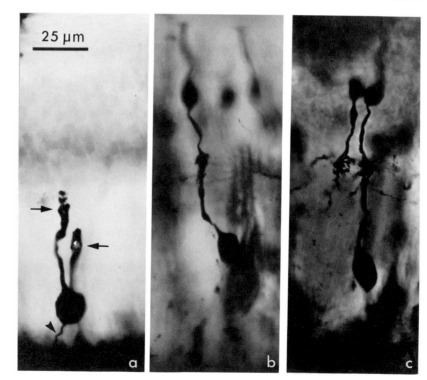

25 µm

ramify in the upper part of the layer. This arrangement implies that the two types of midget bipolar cells described earlier pass on their information to two types of midget ganglion cells. Thus, in the central part of the primate retina one cone sends two separate messages to the brain via two midget ganglion cells. And, as we shall see, one cell most likely signals an increase in illumination (that is, gives an on-response), while the other signals a decrease in illumination (that is, give an off-response).

Many of the features of the cellular organization of the vertebrate retina are shown in Figure 2.12, which is a composite drawing by Cajal (Ramón y Cajal, 1911) of many of the cells found in the frog retina. Various types of receptors, bipolar, amacrine, and ganglion cells are illustrated, along with one partially stained horizontal cell (H) (no axon terminal expansion) and perhaps part of an interplexiform cell (I). Stratification in the inner plexiform layer can be seen readily, as well as two displaced neurons, a displaced amacrine cell (DA), and a displaced bipolar cell (DB). Finally, two pigment epithelial cells (PE) are illustrated, with their long processes that fit between the photoreceptors cells. The apical part of the cells, along with the processes, are filled with light-absorbing black pigment granules that serve to shield the photoreceptors from stray light.

Retinal Processing

What processing occurs within the retina? What are the retinal ganglion cells signaling to the brain? I have already suggested that ganglion cells signal brightness and darkness information, but they also communicate to the brain much more than that. Indeed, there appear to be two basic types of processing occurring in the retina: one carried out mainly in the outer plexiform layer and the other carried out in the inner plexiform layer. Ganglion cells conveying information that reflects the two basic types of processing are known from studies on a variety of retinas and will be described here first in an idealized and schematic way. Later I shall discuss the variety of ganglion cell responses and the thorny problem of the classification of ganglion cells based on their physiological responses.

Ganglion cells typically respond to illumination of a restricted but relatively large region of the retina. This region is called the receptive field of the cell and typically is about 1 mm in diameter (Hartline, 1938; Kuffler, 1953). This arrangement means, of

2.12 Drawing showing typical retinal cells of the frog and the cellular organization of the frog retina. PE, pigment epithelial cells; I, perhaps a partially stained interplexiform cell; DB, displaced bipolar cell; DA, displaced amacrine cell. The arrow points to a so-called Landolt Club process that extends distally from many bipolar cells in cold-blooded vertebrates from the outer plexiform layer to the outer limiting membrane. The function of the Landolt Club is not known in any retina. Modified from Ramón y Cajal (1911), with permission of the Instituto Ramón y Cajal.

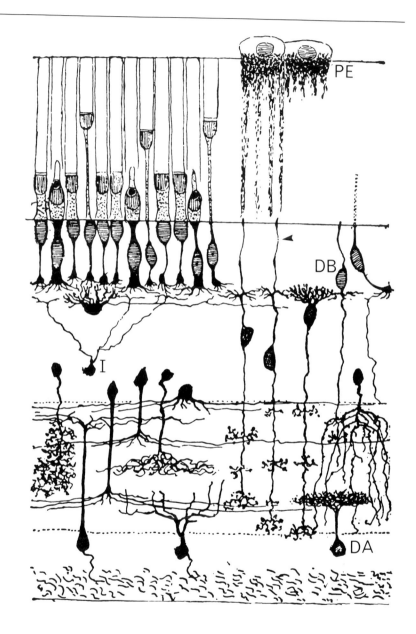

course, that many receptors contribute to the responses of a single ganglion cell as do, in most cases, a number of each of the types of inner nuclear layer neurons. The receptive fields of adjacent ganglion cells overlap considerably, but not completely, so any one region of the retina is covered by many different ganglion cells, each of which is concerned with a slightly different part of the visual field. This arrangement means also that any one receptor, and inner nuclear layer cell, usually contributes to the responses of a number of ganglion cells. (The obvious exception here is the midget bipolar cell, which may provide input to only a single midget ganglion cell.)

Ganglion cells receive direct input from two types of inner nuclear layer cells: bipolar cells and amacrine cells. It appears, furthermore, that the ganglion cell responses strongly reflect the properties of the input neurons; and it has been shown that the properties of bipolar cells and amacrine cells are usually distinctly different. Bipolar cells give sustained responses to retinal illumination, whereas many amacrine cells respond with transient potentials, typically at both onset and cessation of illumination. Some ganglion cells appear to receive most of their input from bipolar cells; their responses are sustained and reflect primarily the processing of information occurring in the outer plexiform layer. Other ganglion cells receive most, if not all, of their input from amacrine cells; their responses are often more transient and reflect inner plexiform layer processing. Although this is a simplified view, I shall show in succeeding chapters that there is anatomical, physiological, and pharmacological evidence to support it.

As mentioned above, the receptive field of one type of ganglion cells reflects the processing occurring in the outer plexiform layer of the retina. The receptive fields of these cells show evidence of spatial processing of visual information (Figure 2.13). Ganglion cells of this type are called contrast-sensitive units and are subdivided into two, mirror-image classes: on-center cells and off-center cells. Such cells, first described in detail by Kuffler (1953) in the cat retina, display a receptive field that is organized in two concentric regions that are antagonistic to each other.

For an on-center cell, an increase in the illumination of the center of the field with a spot of light causes the cell to fire a sustained burst of impulses for as long as the light is on, the frequency of which depends on light intensity. An increase of illumination within the antagonistic surround of the field causes inhibition of the cell while the light is on, as can be seen by a decrease in the maintained activity of the cell, and a vigorous burst of impulses when the light turns off. If both the center and surround are illuminated simultaneously, the two regions antagonize each other, and usually only a weak response, characteristic of the center, occurs. Thus, these cells respond vigorously when illumination is confined to a particular region of the receptive field; they are clearly interested in the spatial

2.13   Idealized responses and receptive field maps for on-center (*top*) and off-center (*bottom*) contrast-sensitive ganglion cells. The drawings on the left represent hypothetical responses to a spot of light presented in the center of the receptive field, in the surround of the receptive field, or in both the center and surround regions of the receptive field. A + symbol on the receptive field map indicates an increase in the firing rate of the cell, that is, excitation; a − symbol indicates a decrease in the firing rate, that is, inhibition.

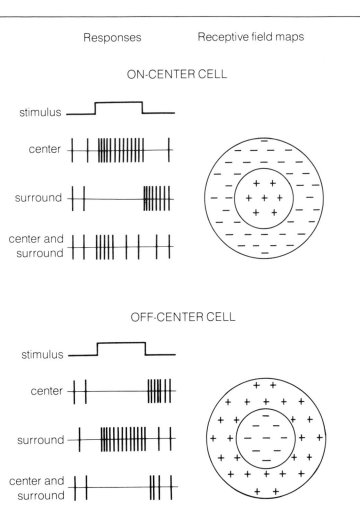

distribution of light on the receptor mosaic. Furthermore, because the center and surround regions are antagonistic, these cells respond even better when there is maximum contrast between the center and surround stimuli, that is, bright center spot and a dark annulus (ring).

The receptive field of the off-center cell is the mirror-image of that of the on-center cell. In other words, an increase of illumination of the center causes an inhibition of the maintained activity of the cell, and increased illumination of the surround causes an increase of activity. Again the cell responds most dramatically when there is maximum contrast of the center and surround regions.

The second basic type of ganglion cells reflects inner plexiform

2.14 Idealized responses and a receptive field map for a direction-sensitive ganglion cell. Such cells respond with a burst of impulses at both the onset and termination of a spot of light presented anywhere in the cell's receptive field. This response is indicated by ± symbols all over the map. Movement of a spot of light through the receptive field in the preferred direction (*open circles*) elicits firing from the cell that lasts for as long as the spot is within the field. Movement of a spot of light in the opposite (null) direction (*open squares*) causes inhibition of the cell's maintained activity for as long as the spot is within the receptive field.

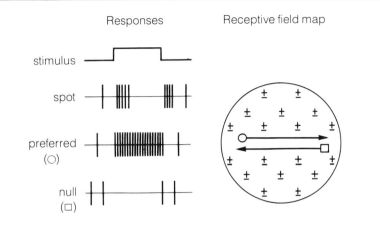

layer responses and shows more evidence of temporal processing of visual information (Figure 2.14). The responses of these ganglion cells to spots of light resemble the responses of the transient amacrine cells. These ganglion cells are usually highly movement-sensitive and many show direction-sensitive responses. They respond vigorously to spots of light moving through the receptive field in a particular direction, but often they respond not at all, or are even inhibited by the same spot of light moving in the opposite direction. Such cells were first described in detail by Barlow et al. (1964) in the rabbit retina. When random areas of the receptive field are illuminated with static spots of light, these cells usually respond with a burst of impulses at the onset of illumination and another burst of impulses at its cessation; thus, they are commonly referred to as on–off ganglion cells as well as movement- or direction-sensitive ganglion cells. Because the on–off responses of these cells do not usually depend on where in the receptive field a spot of light is presented, these cells are not so concerned with the spatial aspects of illumination.* Furthermore they usually respond as well to a dark spot moving across an illuminated receptive field as to a bright spot moving on a dark field. Because they are strongly acti-

---

* When a spot, either stationary or moving, is presented to the retina just beyond the receptive field of the cell, no responses occur either during or after the stimulus. However, such peripheral illumination often will depress the response of the cell to a stationary or moving stimulus presented anywhere in the receptive field (Barlow et al., 1964). This phenomenon is sometimes referred to as a silent inhibitory surround. Evidence suggests that this inhibitory surround is simply an extension of the inhibition that underlies the generation of the direction-sensitive receptive field of these cells (Wyatt and Daw, 1975; see also Chapters 4 and 5).

vated by changes in illumination—its onset or cessation or move-ment—they are clearly more responsive to the temporal aspects of the stimulus.

Virtually all retinas appear to have ganglion cells of both types. Some retinas, however, appear to have more ganglion cells of one type than the other. For example, cats and monkeys have many sustained on- and off-center ganglion cells and few on–off, direc-tion-sensitive ganglion cells; conversely, many cold-blooded verte-brates have many transient on–off ganglion cells and relatively few sustained on- and off-center cells. These differences between species will be discussed further in Chapter 3.

It is also true that a variety of other types of ganglion cells have been distinguished on the basis of receptive field organization. Some of the receptive-field responses can be satisfactorily explained by postulating simple mixes of bipolar and amacrine cell inputs into the ganglion cells or variations in the response properties of the input neurons. For example, in many retinas there are transient on, direction-sensitive ganglion cells that one would presume re-ceive their input mainly from transient amacrine cells that respond only at the onset of illumination (and such amacrine cells have been observed). In a number of retinas, some ganglion cells have recep-tive fields that are organized in a concentric, on- or off-center fash-ion, but the light responses of these cells are much more transient to patterns of light than are those of the sustained on- and off-center ganglion cells shown in Figure 2.13. The presumption here is that these cells get a mix of transient amacrine and sustained bipolar cell input. Finally, some ganglion cells have receptive field properties and organization that are not explained so easily, al-though ideas concerning their generation can be suggested. These notions will be taken up at the end of Chapter 4.

Classification of Ganglion Cell Receptive Fields

Hartline (1938) was the first to record from single ganglion cell axons and to classify ganglion cells on the basis of their receptive field properties. Most of his experiments were in the frog, but he also recorded from some ganglion cells in the mudpuppy, turtle, alligator, and shark. He found that the responses generally fell into three categories: on cells, off cells, and on–off cells. He noted that both on-center and off-center cells tended to give more or less sus-tained responses when the light was on, whereas the on–off cells gave very transient responses at the onset and cessation of light. He noted that in some animals there were variations of these basic response types: "in both the turtle and alligator eyes, fibers are not infrequently found giving only a brief burst when light is turned on, with neither maintained discharge nor an 'off' response" (Hartline, 1938, p. 410).

In 1953 Barlow and Kuffler reported that the receptive fields of the on- and off-center cells in frogs and cats, respectively, could be divided into two concentric zones such that illumination of the surround antagonized the center response. In the partially light-adapted cat retina, the center–surround antagonism was shown to be particularly striking. Thus the properties of these fields have been extensively studied since.

In Barlow's paper of 1953, he also pointed out that the frog on–off ganglion cells are highly sensitive to moving stimuli and seem ideally suited to explain the fact that, when feeding, frogs respond very well to small moving objects but poorly or not at all to stationary objects. Thus, he suggested that the on–off cells can be thought of as "fly-detectors." This notion that different ganglion cell types have specific functional roles was taken up by Maturana and Lettvin and their colleagues in the late 1950s. By reexamining frog ganglion cells from this point of view, they differentiated ganglion cell responses in terms of the stimulus that appeared most effective for activating the cell. The cells were named according to the operation they appeared to be carrying out: boundary detectors; movement-gated, dark convex boundary detectors; movement or changing-contrast detectors; dimming detectors; and unclassified cells (Maturana et al., 1960). This idea of classifying cells on the basis of what Barlow called the "trigger feature" of a cell type, rather than on their responses to stationary spots of light, was extended in a number of studies. So, for example, when Barlow et al. (1964) discovered that the on–off ganglion cells in the rabbit retina responded to movement in a particular direction, they named the cells "directionally selective" cells.

In addition to the on–off direction-sensitive cells, Barlow et al. (1964) described four other types of retinal ganglion cell receptive fields in the rabbit: on-center, off-center, on direction-sensitive, and large-field units (cells that had unusually large receptive fields and gave vigorous responses to fast-moving bright spots on a dim background or dark spots on a bright background). Subsequently, Levick (1967) described three additional classes of ganglion cells in the rabbit: orientation-selective cells, local-edge detectors, and uniformity detectors.

In 1966, Enroth-Cugell and Robson discovered that on-center and off-center cells in the cat could be divided into two types on the basis of their responses to sinusoidal gratings; these two types were given the noncommittal names X and Y.* The X-cells were found to give much more sustained responses than do the Y-cells, which often gave only a brief burst of impulses at the onset or cessation of illumination (see Cleland et al., 1971). The X-cells behave very much like the sustained on- and off-center cells described in Figure 2.13, whereas the Y-cells behave as though they get a mix of sustained (bipolar) and transient (amacrine) input (Victor and Shapley, 1979). The ganglion cells giving rise to the X-type and Y-type responses have now been identified morphologically; and it has been shown that the X-cells make up about 55 percent of the ganglion cell population in the cat, whereas the Y-cells make up just 4 percent (Wässle et al., 1981a,b).

The remaining 40 percent of the ganglion cells in the cat retina are not nearly as well categorized, but many of them seem to have receptive field properties similar to many of the rabbit ganglion cells (Stone and Fabian, 1966; Stone and Hoffman, 1972; Cleland and Levick, 1974b). Thus, on–off units were found, some being direction-sensitive cells and some local-edge detectors. Other cells respond to light sluggishly, although they can be mapped as on- or off-center units. All of these cells in the cat retina appeared to have relatively small cell perikarya (presumably why they are recorded infrequently and are not well studied) and were classified together as W-cells (Stone and Fukuda, 1974).

There is no general consensus concerning ganglion cell classification (Stone, 1983). Some investigators use the X, Y, W classification scheme, others use a trigger-feature classification scheme, and still others use a combination of these two. Because both of these schemes reflect to some extent the properties of the cells in individual species (for example, the Y-cells of the cat appear in

---

* Enroth-Cugell and Robson (1966) observed that a sinusoidal grating can be positioned on the receptive fields of some ganglion cells (X-cells) in such a way that no response occurs when the grating is introduced onto, or withdrawn from, background illumination. Other cells (Y-cells) always respond when the grating is introduced or withdrawn under similar conditions. A null position can be found for X-cells because these cells respond in a symmetrical way to the onset and offset of illumination. In other words, their on and off components are essentially mirror images of one another; hence, a position in the receptive field can be found such that adding and subtracting equal amounts of illumination from comparable regions of the receptive field elicits no response. Put another way, X-cells summate responses to illumination of different parts of the receptive field in a linear fashion, but Y-cells summate such responses nonlinearly.

many respects to be unique to the cat), it seems to me that it is desirable to find ways to classify ganglion cell responses in a more universal way. In this book, I emphasize the mechanisms by which the ganglion cell receptive fields are formed, principally from the types of inputs they receive. It is possible that eventually this will be a more appropriate way to classify ganglion cells. At the moment, the kinds of input a ganglion cell receives usually cannot be determined, especially from extracellular recordings, but methods are being developed to do this (see, for example, Victor and Shapley, 1979; K.-I. Naka, personal communication).

THREE

Wiring of the Retina

AN important first step in attaining an understanding of any part of the brain, including the retina, is to learn how the cells connect together. Nerve cells interact at discrete and specific sites, called synapses, and knowledge of the structure of the synaptic junctions and of the "wiring" of a neural system obviously will provide important clues to functional mechanisms.

Light microscopy of retinas stained by the Golgi method or similar methods shows the distribution of processes within a piece of neural tissue and provides information about possible interactions. For example, we know from light microscopic studies that the processes of four cell types are found in each of the two plexiform layers in the retina: processes from the receptor, horizontal, bipolar, and interplexiform cells are present in the outer plexiform layer; and processes from bipolar, amacrine, interplexiform, and ganglion cells are present in the inner plexiform layer. If the main direction of information flow through a system is known—as it is for the retina, that is, receptors to ganglion cells—we can infer likely synaptic contacts from light microscopic studies. Consequently, we conclude that receptor terminals are likely to make synaptic contacts onto bipolar and horizontal cell processes, whereas bipolar terminals probably make synaptic contacts onto amacrine cell processes and ganglion cell dendrites, especially because these terminals and cell processes can be seen to closely approach one another.

For a detailed view of the synapses or insight into the patterns of synaptic connections between neurons, electron microscopy is needed to provide sufficient resolution of cell structures. Indeed, the information yielded by electron microscopic studies provides clear answers to old puzzles. For example, since Cajal's discovery more than 100 years ago of axonless (amacrine) cells in the retina and elsewhere, numerous investigators have wondered how such cells might function with only "dendrites." Electron micrographs showed unambiguously that amacrine cell processes have morphological characteristics of both dendrites and axons (Dowling and Boycott, 1966). They can both receive and make synaptic junctions, a characteristic that makes amacrine cells a versatile kind of neurons to be involved in mediating interactions within a plexiform layer (see below and Chapter 8).

Before an effective electron microscopic analysis of a piece of neural tissue can be carried out, two major difficulties must be resolved: (1) anatomical recognition of synaptic junctions and (2) identification of the cell types and their processes in the electron microscope. Even today these difficult problems have not been completely resolved. Nevertheless, certain specializations observed at known synaptic junctions are not seen at nonsynaptic sites in the vertebrate nervous system, and these specializations are consistently observed throughout the nervous system (Palay, 1956; Gray, 1959). It also is possible to relate a number of the specializations with known components of the synaptic transmission process (Pappas and Waxman, 1972). Thus, the correlation between a recognizable anatomical structure and a point of synaptic contact in many parts of the vertebrate nervous system is secure. In the retina we observe such synaptic junctions, and with some confidence we label these as synapses. Indeed, we believe we see the sites of both chemical and electrical synapses between retinal neurons. We also have good physiological evidence for both chemical and electrical synaptic transmission in the retina (see Chapters 4 and 5), and the two approaches are, for the most part, very complementary.

We also have in the retina, particularly in the outer plexiform layer, some arrangements of processes at presumed synapses that are not seen elsewhere in the nervous system and junctions that do not fulfill all or even many of the criteria for synaptic junctions. In some cases these unspecialized junctions are thought to be synaptic. Furthermore, we know of at least one specific retinal interaction from physiological studies—feedback from horizontal cells to cone terminals—for which no anatomical correlate has been identified, even though such a junction has been searched for extensively. Thus, questions remain with regard to the synapses and synaptic organization of the outer plexiform layer. But much has been learned about the fine structure of the outer plexiform layer, and a number of educated guesses concerning its synaptic organization can be made.

In the inner plexiform layer the synaptic morphology appears more conventional, and fewer questions concerning the identification of synaptic sites have been raised. The inner plexiform layer of the retina is much more extensive than is the outer plexiform layer and has more synapses per unit area; however, it has been studied in less detail than has the outer plexiform layer. We have only general notions of specific synaptic pathways through the inner plexiform layer, although physiological data and information gleaned from comparative studies on inner plexiform layers from different animals provide insights about the functional mechanisms of this layer.

The difficulty of identifying specific cell types and their processes in the electron microscope is, in a sense, a less challenging problem than that of identifying the synaptic junctions, primarily because the former is more technical in nature. It is possible in theory to trace all cellular processes back to their cell body of origin by use of serial sections and to identify the cells and their processes by shape and position in the retina. A simple calculation, however, shows that this is not a practical approach. Reconstruction from electron micrographs of just one cubic millimeter of tissue magnified 10^4 times would require approximately 10^8 micrographs. Thus, serially sectioning through an entire retina is not reasonable, although successful efforts at serially sectioning small pieces of the retina have been made (Kolb, 1979; Stevens et al., 1980).

Nevertheless, there are ways to identify classes of cells and their processes in the electron microscope without serial sectioning. Various classes, and even parts of nerve cells, differ in their fine structural appearance (Figure 3.1). Axon terminals characteristically contain numerous small vesicles—the synaptic vesicles—which are not found in dendrites. And dendrites frequently contain ribosomal particles that are not found in axon terminals. Furthermore, the terminals of different cell classes may make different kinds of synaptic contacts (which is the case in the retina). Thus, morphological differences, coupled with the fact that in each plexiform layer the processes of just four cell classes interact, make it possible to identify routinely the kind of cell or process one is viewing.

A number of specialized methods also can be used to identify unequivocally classes, types, and even subtypes of cells in electron micrographs. Viewing Golgi-stained cells in the electron microscope enables one to determine securely the identification of processes, particularly those on the postsynaptic side of synapses (see Stell, 1965; inset in Figure 3.6). Substances such as horseradish peroxidase can be injected into individual cells that have been recorded intracellularly. Subsequently, these cells can be identified and studied in the electron microscope and their synaptic contacts determined (Plate 4). Specific histochemical or immunohistochem-

3.1 **a** Low-power electron micrograph of the outer plexiform layer of the cat retina. The vesicle-filled receptor terminals (RT) are aligned along the distal margin of the outer plexiform layer, whereas a laterally extending horizontal cell (H, *dashed line*) and several bipolar cell perikarya (B) are found along the inner margin of the plexiform layer. The larger receptor terminals are cone endings; the smaller spherical terminals are rod endings. V, blood vessel. From Dowling et al. (1966), reprinted with permission; copyright 1966 by the AAAS.

b Low-power electron micrograph of a portion of the inner plexiform layer of the human retina, showing two large, bulbous bipolar terminals (BT). The bipolar cell terminals contain numerous synaptic vesicles and occasional synaptic ribbons (*arrowheads*). The smaller processes in the micrograph are mainly amacrine cell processes and ganglion cell dendrites. From Dowling (1970), reprinted with permission of J. B. Lippincott Company.

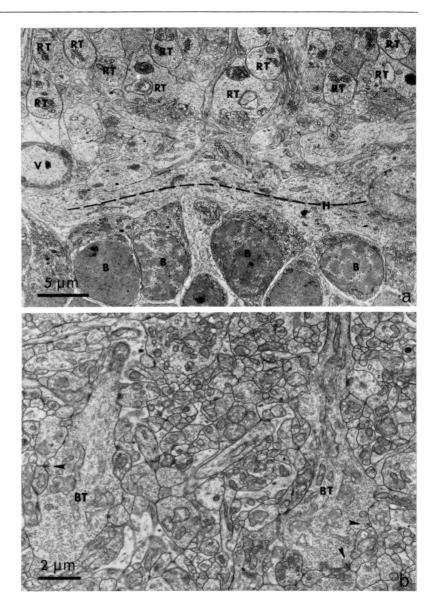

ical staining of neurons can also be used to identify specific types and subtypes of cells in the electron microscope and to gain information concerning the synaptic organization of individual cells and chemical-specific pathways (see Figure 3.12). These three methods and others are providing a wealth of information.

Retinal Synapses

In both plexiform layers two types of specialized contacts commonly observed are believed to represent chemical synaptic junctions (Kidd, 1962; Missotten, 1965; Dowling and Boycott, 1966; Raviola and Raviola, 1966). One of these is quite similar in morphology to known chemical synapses seen throughout the vertebrate nervous system and thus is called a conventional synapse. In the retina horizontal, amacrine, and interplexiform cells and, when present, centrifugal fibers make conventional synapses. The other type of synapse is characterized by an electron-dense ribbon or bar in the presynaptic process and is called a ribbon synapse. The terminals of a variety of sensory cells make ribbon synapses (Smith and Sjöstrand, 1961; Flock, 1964) and similar synapses are observed throughout many invertebrate nervous systems (Dowling and Chappell, 1972; Muller and McMahon, 1976; Armett-Kibel et al., 1977). In the retina ribbon synapses are made by the receptor and bipolar cells. There is, in addition, a junction made by the receptor terminals, mainly with bipolar cells, that is relatively unspecialized and that has not been described elsewhere in the nervous system. These contacts were first observed on the flattened, basal surface of mammalian cone terminals and hence were called superficial, or flat, or basal junctions. We will call them basal junctions (Lasansky, 1969). Because certain bipolar cells make contacts with receptors only at basal junctional sites, it is generally believed that these contacts also are sites of chemical synaptic interactions.

Junctions indicative of electrical synaptic contacts between cells also are seen in both plexiform layers of the retina. These synapses appear similar to the electrical junctions observed between neurons elsewhere (Bennett et al., 1963; Pappas and Waxman, 1972) and, as elsewhere, are called gap junctions. In the retina gap junctions are frequently seen between receptor terminals, horizontal cells, and amacrine cells (Witkovsky et al., 1974, 1983; Naka and Christensen, 1981) and occasionally between bipolar terminals or between bipolar terminals and amacrine cell processes (Witkovsky and Stell, 1973; Famiglietti and Kolb, 1975).

Conventional Synapses

As already noted, conventional synapses in the retina resemble chemical synapses seen throughout the vertebrate nervous system (Pappas and Waxman, 1972). They are characterized by an aggregation of synaptic vesicles in the presynaptic terminal clustered close to the presumed presynaptic membrane (Figure 3.2). Some electron-dense material frequently is observed in association with the presynaptic membrane and probably corresponds to the electron-dense projections seen on the inner surface of the presynaptic membrane of many synapses. This electron-dense material is believed to play a role in the attachment of the synaptic vesicles to the presynaptic membrane.

Some filamentous material often can be seen in the synaptic cleft. But in the retina the cleft generally is not widened at conventional synapses as it is at many chemical synapses in the vertebrate nervous system. At conventional synapses in the retina the synaptic cleft is typically ~20 nm wide, whereas at many conventional synapses seen elsewhere in the brain the synaptic cleft is ~30 nm wide. Usually some electron-dense material is associated with the inner surface of the postsynaptic membrane, but the amount and density of this staining usually is not as prominent as that seen at many vertebrate synapses.

At many synapses in various places in the vertebrate brain, the synaptic vesicles may appear flattened when the tissue is appropriately fixed (Uchizono, 1967). The junctions with flattened vesicles are thought to represent inhibitory synapses; other terminals in the same material have round synaptic vesicles, however, and are believed to represent excitatory synapses. In the retina such a dichotomy of synaptic vesicle shapes has not been observed, regardless of the method of fixation used; generally the synaptic vesicles observed are more or less round. Most conventional synapses

3.2 Drawing and electron micrograph of a typical conventional synapse observed in the vertebrate retina (frog inner plexiform layer). These synapses are characterized by an aggregation of synaptic vesicles clustered close to the presynaptic membrane and some electron-dense material associated with both the pre- and postsynaptic membranes. Fine filamentous material can be distinguished crossing the synaptic cleft. Modified from Dowling (1968), with permission of the Royal Society.

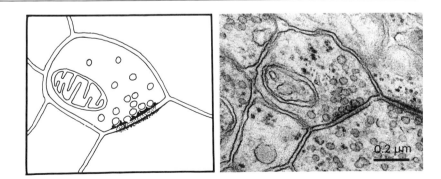

in the retina are thought to be inhibitory in nature, because they are made mainly by the horizontal and amacrine cells, which are known to be mainly inhibitory (Chapter 5). Why the vesicles of the retinal synapses do not show these shape differences upon appropriate fixation is not understood.

Ribbon Synapses

Ribbon synapses are characterized by an electron-dense ribbon or lamella in the presynaptic cytoplasm (Sjöstrand, 1953; Ladman, 1958; Gray and Pease, 1971). Typically the ribbon is oriented at right angles to the presynaptic membrane and has some depth (~1 μm). It usually sits in or just above an evaginated ridge of the terminal membrane (Figure 3.3). Between the ribbon and the ridge membrane is a curved dense band, called the arciform density, which may serve to anchor the ribbon to the membrane. Surrounding the synaptic ribbon is a precisely arranged array of synaptic vesicles, and high-resolution electron microscopy often shows thin filaments extending from the ribbon to the synaptic vesicles. Curiously, the ribbons observed in the outer plexiform layer are usually significantly longer than those in the inner plexiform layer.

Apposed to the synaptic ridge are multiple processes. In the inner plexiform layer, typically two processes (a dyad) are seen; in the outer plexiform layer, three (a triad) are often observed. Usually electron-dense material is associated with the membranes of the processes apposed to the ridge containing the synaptic ribbon—but not always. The cleft between ridge and apposed processes is widened, and filamentous material runs between the ridge and the apposed processes. These findings imply that, at ribbon synapses, multiple postsynaptic processes receive input.

Recent electron microscopic studies of ribbon synapses in the outer plexiform layer using the method of freeze-fracturing have shown that on either side of the ridge synaptic vesicles bind to the membrane and presumably release transmitter at those sites (Raviola and Gilula, 1975). These findings suggest that the ribbon serves as a conduit for the synaptic vesicles to the release sites, much as the electron-dense projections along the presynaptic membranes of conventional synapses serve to aid in the binding of synaptic vesicles to the presynaptic membrane. In agreement with this suggestion, no electron-dense projections have been observed along the synaptic ridge of ribbon synapses.

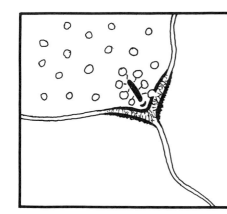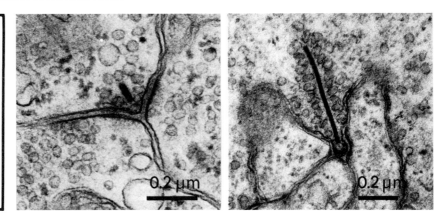

3.3 Drawing and electron micrographs of retinal ribbon synapses in the frog. The ribbon is surrounded by a halo of synaptic vesicles. The ribbon synapse of the inner plexiform layer (*drawing and left micrograph*) has two postsynaptic processes; the ribbon synapse of the outer plexiform layer (*right micrograph*) has three. Electron-dense material is associated with the membranes of both pre- and postsynaptic processes. Note that the ribbon in the photoreceptor terminal in the outer plexiform layer is much longer than that in the bipolar cell terminal in the inner plexiform layer. Modified from Dowling (1968), with permission of the Royal Society.

Basal Junctions

Basal junctions are made by the receptor terminals. In mammals these junctions are made only by cone cells, although in a number of nonmammalian vertebrates they are observed in both rod and cone terminals. Typically, at a basal contact the membrane of the receptor terminal is smoothly indented and there is prominent electron-dense material on the inner surface of the terminal membrane (Figure 3.4). The cleft between the terminal and the contacting process is widened, and filamentous material often can be seen within the cleft. No vesicles are associated with the contact area; indeed, at many basal junctions there appears to be a clear, vesicle-free zone immediately adjacent to the terminal membrane along the area of contact. Usually little, if any, electron-dense material is associated with the membrane of the process contacting the receptor, so in most cases the junction has an asymmetrical appearance.

These junctions are believed to be synaptic because certain bipolar cells (the flat bipolar cells) have only this kind of junction with the receptor (see below). The absence of synaptic vesicles as-

3.4 Drawing and electron micrograph of a typical basal junction made primarily by cone receptor terminals. Electron-dense material on the receptor membrane is apparent, as well as filamentous material in the cleft between the terminal and contacting process. The micrograph is from the frog. Modified from Dowling (1968), reprinted with permission of the Royal Society.

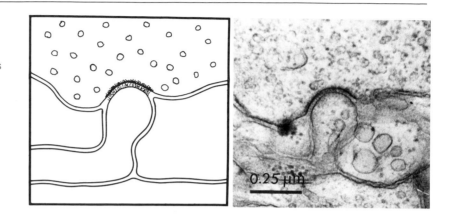

sociated with these contacts is a puzzle and raises questions as to how these junctions function. An obvious suggestion is that at these junctions synaptic transmitter is not released from vesicles. This hypothesis is supported by freeze-fracture analyses, which have failed to reveal any evidence of vesicle binding to or vesicle release of transmitter from the receptor membrane at these junctions (Raviola and Gilula, 1975). On the other hand, freeze-fracture studies have shown particles that are contained within the membranes of the contacting bipolar cell processes and are similar to particles typically associated with postsynaptic membranes at excitatory synapses seen in many parts of the brain. These latter findings support the notion that these junctions are chemical synapses.

Some workers have noted that the cleft at certain basal junctions is narrower than those at other basal junctions and that the amount and distribution of electron-dense material associated with the cell membranes at the point of contact varies (Lasansky, 1971; Stell, 1978). They have suggested that there are subtypes of basal junctions, for example, wide-cleft and narrow-cleft junctions. Freeze-fracture analyses have not revealed differences between these varieties of basal junctions, however; a fact suggesting that functionally there is only a single type of basal junction (Schaeffer et al., 1982).

Gap Junctions

Retinal gap junctions are very similar in their morphology to electrical junctions observed in other neural tissue (Pappas and Waxman, 1972; Raviola and Gilula, 1973, 1975). The membranes of the contacting cells come into close apposition, being separated by

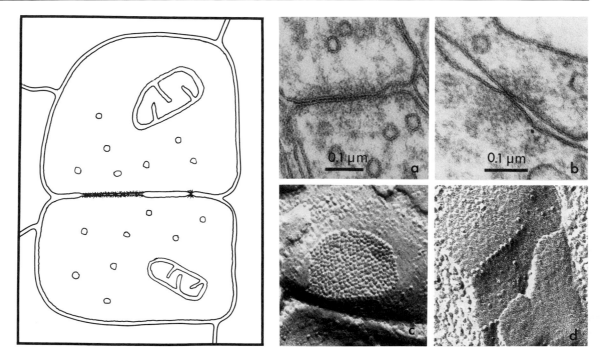

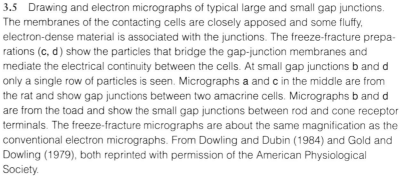

3.5 Drawing and electron micrographs of typical large and small gap junctions. The membranes of the contacting cells are closely apposed and some fluffy, electron-dense material is associated with the junctions. The freeze-fracture preparations (c, d) show the particles that bridge the gap-junction membranes and mediate the electrical continuity between the cells. At small gap junctions **b** and **d** only a single row of particles is seen. Micrographs **a** and **c** in the middle are from the rat and show gap junctions between two amacrine cells. Micrographs **b** and **d** are from the toad and show the small gap junctions between rod and cone receptor terminals. The freeze-fracture micrographs are about the same magnification as the conventional electron micrographs. From Dowling and Dubin (1984) and Gold and Dowling (1979), both reprinted with permission of the American Physiological Society.

a gap of no more than 2–4 nm (Figure 3.5). The gap between the cells usually is not continuous, but typically is interrupted by bridging structures. Some fluffy, electron-dense material characteristically is associated with the gap junctional membranes, and the apposing membranes typically appear very regular and distinct at the junction.

The nature of the bridging structures is made clear by freeze-fracture of the junctions, which reveals very prominent particles within the membranes at the sites of contact. The particles of contacting cells are thought to be apposed across the extracellular gap and to form channels enabling ions and small molecules (up to a molecular mass of about 1,000 daltons) to flow directly from one cell to another. The flow of substances from one cell to another can be readily demonstrated by injecting fluorescent dyes into one cell and visualizing dye passage into adjoining cells (Plate 3) (Kaneko, 1971a).

In the retina there is considerable variety in the extent of gap junctional area between adjacent cells and also in the distribution of the intramembranous particles at various gap junctions. For example, the gap junctions observed between horizontal cells, especially in cold-blooded vertebrates, are especially large (see Figure 3.11), whereas some observed between photoreceptor cells are tiny (Figure 3.5). Indeed, the latter junctions, sometimes called kissing junctions, seem to consist of a single row of connecting particles (Raviola and Gilula, 1973). There is evidence that the extent of coupling depends on the size of the junction (Gold and Dowling, 1979), but it has also been observed that the density of connecting particles can differ enormously between gap junctions (compare Figures 3.5 and 3.11). Furthermore, the conductance of the gap junctions between horizontal cells can be modulated by synaptic input to the cells (Lasater and Dowling, 1985a,b); so the relation between strength of coupling and gap junction size can be complicated.

Other Junctions

Other specialized contacts observed in the retina, particularly in the outer plexiform layer, may have significance for neuronal communication. In the outer plexiform layer of the primate, numerous specialized attachments (or desmosomal-like junctions) are seen (Dowling and Boycott, 1966). Recent freeze-fracture data indicate that these junctions are quite different from desmosomes and appear to be a unique type of contact (Raviola and Gilula, 1975). Lasansky (1971, 1973) has also noted various specialized contacts in the outer plexiform layer of both the turtle and the salamander. As yet no one has presented convincing arguments that any of these junctions are synaptic.

Synaptic Organization

In each plexiform layer, usually just three cell classes form chemical synapses. The input neurons—the receptors and bipolar cells—

form ribbon synapses, and horizontal, amacrine, and interplexiform cells form conventional synapses. The dendrites of the output neurons in the two plexiform layers, the bipolar cells and the ganglion cells, do not appear in most cases to make chemical synapses; they are usually only postsynaptic (but see Sakai et al., 1986). By far the most complex synaptic arrangements involve the receptor terminals.

Outer Plexiform Layer

PHOTORECEPTOR CELLS   In most retinas the receptor synaptic ribbons are associated with prominent invaginations present along the base of the receptor terminal. Processes from the outer plexiform layer penetrate into the invaginations and terminate near the ridge in which the synaptic ribbon lies (Figure 3.6). The arrangements of both the invaginated ribbon synapses and superficial basal junctions is particularly clear in the case of the terminals of the central primate receptors, and these will be our model (Missotten, 1965; Dowling and Boycott, 1966; Kolb, 1970). In primates and other mammals, the cone terminals are larger than the rod terminals; they are often called pedicles whereas the smaller rod terminals are called spherules. Dendritic processes from horizontal cells and three types of bipolar cells (flat bipolar cells and invaginating

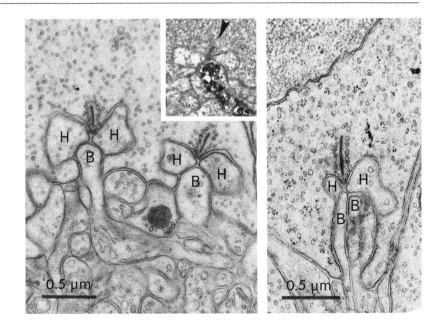

3.6   Electron micrographs of the invaginated ribbon synapses of receptor terminals. In the cone terminal (*left*), three processes penetrate into each invagination; in the rod terminal (*right*), four or more processes are observed within the invagination. The processes that lie laterally and deeper in the invaginations are horizontal cell processes (H); the central elements are bipolar cell dendrites (B). *Inset,* a Golgi-stained dendrite from an invaginating midget bipolar cell terminating as the central element in a cone terminal. Arrowhead points to the synaptic ribbon. Left micrographs are of monkey foveal cone terminals; right, of a cat rod terminal. From Dowling (1970), reprinted with the permission of J. B. Lippincott Company.

and flat midget bipolar cells, see Figure 3.7) contact the central cone pedicles in primates. The dendrites of the horizontal cells and the invaginating midget bipolar cells penetrate into the invagination, whereas the dendrites of the flat midget bipolars terminate at basal junctions on the base of the terminals. Invariably two horizontal cell processes and one midget bipolar dendrite enter each invagination. The horizontal cell processes extend somewhat deeper into the invagination and end lateral to the synaptic ribbon ridge, whereas the invaginating midget bipolar dendrite lies centrally within the invagination and somewhat away from the synaptic ridge.

Each cone pedicle has numerous invaginations. Even in the most central region of the retina where the pedicles are smaller, as many as fifteen to twenty-five invaginations may be present in each terminal. In one instance Kolb (1970) serially sectioned a primate cone pedicle contacted by an invaginating midget bipolar cell that had been impregnated with silver by the Golgi method (inset, Figure 3.6). The cone terminal contained twenty-five invaginations, and the central process in twenty-four of the invaginations was from the stained midget bipolar cell. Some sections were lost from the series, which might explain why no stained process was observed in the twenty-fifth invagination.* This midget bipolar cell contacted no other receptor terminals, however, a finding confirming the idea that midget bipolars contact only a single cone terminal. Furthermore, these results show clearly that a single bipolar cell receives multiple inputs from the same receptor terminal.

Examination of other Golgi-stained cells confirmed that the lateral elements within the primate cone pedicle invaginations are always horizontal cell processes. These studies also showed that the flat bipolar cell dendrites, which typically contact six to eight terminals, terminate at basal junctions somewhat away from the invaginations. (As noted earlier, basal junctions only slightly indent the receptor base; and because the base of the receptor terminals in primates is usually very flat, the appearance of the terminal dendritic arborization is characteristically flat, hence the name of these bipolar cells; see Chapter 2.) Finally, the studies showed that the flat midget bipolar cells contact only a single cone and that their

---

* An alternative explanation is that the central process of this invagination was from a diffuse invaginating bipolar cell. Such bipolar cells have been observed in the central regions of the primate retina (Mariani, 1981). One bipolar cell analyzed by electron microscopy contacted seven cone terminals but made as few as two contacts with a single cone terminal (Mariani, 1981).

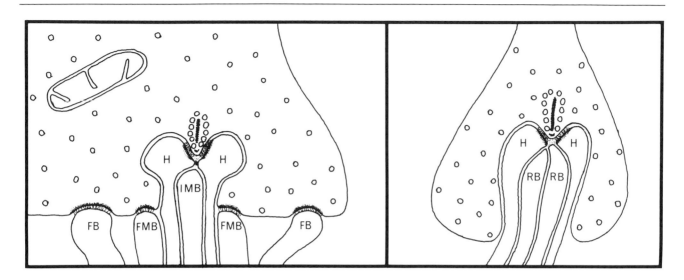

3.7 Schematic drawings showing the arrangements and kinds of junctions made by bipolar cell dendrites and horizontal cell processes with cone (*left*) and rod (*right*) receptor terminals in the primate retina. Horizontal cell processes (H) always end laterally and deeper in the invaginations of both the rod and cone terminals. The invaginating midget bipolar dendrites (IMB) end as the central element in the cone terminal invaginations, whereas the rod bipolar dendrites (RB) end as the central elements in the rod terminal invagination. The flat midget bipolar dendrites (FMB) terminate at basal junctions that are in close proximity to an invagination, whereas the dendrites of the other flat bipolar cells (FB) usually contact the basal surface of the cone terminal somewhat away from the invaginations.

dendrites characteristically terminate superficially on the terminal, but in close proximity to the invaginations. Indeed, a flat midget bipolar appears to contact the receptor on either side of each invagination; consequently, the flat midget bipolars have twice as many junctions with the cone terminals as do the invaginating midget bipolar cells. These relationships are shown in Figure 3.7.

Rod terminals (spherules) in the primate retina have a single invagination into which four or sometimes more processes penetrate (Figures 3.6 and 3.7). The elements that end deeper and lateral to the synaptic ribbon are always processes from the axon terminals of the horizontal cells. The central elements are from different rod bipolar cells. No basal junctions are made by rod terminals in primates.

We still do not understand the significance of the complex organization of the receptor terminal junctions, even in primates where the situation has been very well analyzed and where the synaptic organization seems simpler than in many other animals. There is evidence that the basal junctions of receptors act like classic excitatory synapses. However, the ribbon synapses made between the receptors and invaginating bipolar dendrites can act quite differently. That is, bipolar cells receiving input at ribbon synapses can respond to light with a response the polarity of which is different from that of bipolar cells receiving input at basal junctions (Famiglietti et al., 1977; Stell et al., 1977)—but this is not always the case (Sakai and Naka, 1983; Saito et al., 1983). The receptor–horizontal cell ribbon synapse appears always to be excitatory in nature, but the function of the invaginations is not understood. One speculation is that the invaginations allow for interactions between the horizontal cell processes and the bipolar cell dendrites. The deeply inserted horizontal cell processes are strategically positioned within the invagination to regulate transmitter flow from receptor to bipolar cell dendrite. There is no evidence for this, however. The processes within a receptor invagination are also removed from possible glial cell influence; and this could be a second reason for the invagination of receptor cell synapses.

The organization of the receptor terminals in other species, especially nonmammalian vertebrates, appears less regular and often more complex (Dowling and Werblin, 1969; Lasansky, 1971, 1973). In many species distinct invaginations related to one ribbon complex are not obvious. Instead, a single large invagination encompasses many, if not all, of the processes contacting a particular terminal. In cold-blooded vertebrates most receptor terminals make both ribbon and basal synapses, and both kinds of synapses can be found within these large invaginations. For this reason it is impossible in these species to distinguish bipolars making one or another of these contacts on the basis of the appearance of their dendritic arborization in the light microscope. Figure 3.8 shows a number of these variations in two electron micrographs and a somewhat schematic reconstruction of a terminal from the skate retina (Dowling, 1974).

In some receptor terminals in cold-blooded vertebrates, only two horizontal processes are associated with a ribbon synapse (Lasansky, 1971), whereas in other situations ribbon synapses are observed along the basal surface of the terminal and are not invaginated at all (Dowling and Werblin, 1969). In some instances, numerous processes approach a single synaptic ribbon ridge, and the processes twist and turn within close proximity of the ribbon

3.8 Electron micrographs (**a**) and a summary diagram based on a serial section analysis (**b**) of the receptor terminals in the skate.

Three ribbons are positioned around a single, large invagination. The elements laterally positioned with regard to the synaptic ribbons are horizontal cell processes (H); the centrally placed elements appear to be bipolar dendrites (B_1). A basal junction can be seen within the invagination between the synaptic ribbon complexes (*arrow in inset*). These junctions frequently occur in pairs and are marked by an aggregation of granular material on the presynaptic side of the contact. The postsynaptic elements at these contacts appear to be dendrites from a second class of bipolar cell (B_2). A number of the profiles observed within the invagination are derived from the receptor terminal itself. These processes often appear to surround and to isolate the synaptic junctions. Fine processes from the receptor terminal extend out laterally to contact nearby receptors. Modified from Dowling (1974), with permission of Raven Press, New York.

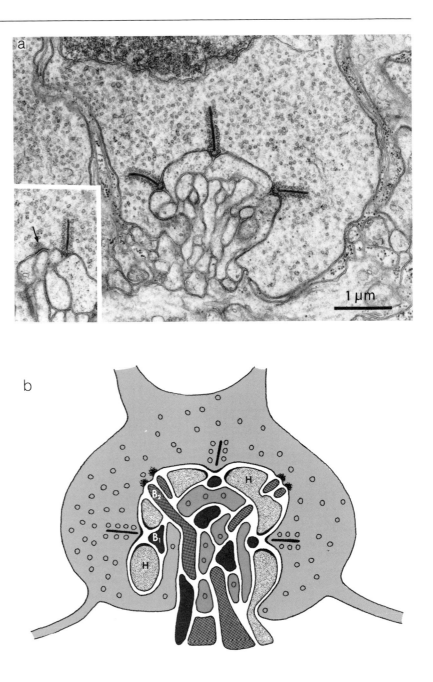

complex, thereby making it difficult to decide what is a lateral process and what is a central process. Finally, a single bipolar cell can be involved with both ribbon and basal junctions in the same terminal (Dacheux, 1982; Saito et al., 1983; Sakai and Naka, 1983). The variability in receptor terminal organization in cold-blooded vertebrates is confusing, but I shall come back to the problem of receptor cell synapses in Chapter 5, when I discuss the generation of bipolar and horizontal cell responses.

In addition to synapses made with horizontal and bipolar cells, photoreceptor cells make junctions with each other (Sjöstrand, 1958). Again a great variety of interreceptor contacts have been seen, but certain generalizations can be made. Most specialized junctions occur between adjacent receptor terminals or between the fine processes, or telodendria, that extend laterally from many receptor terminals (see Figure 2.4) (Cohen, 1965; Lasansky, 1971; Witkovsky et al., 1974; Kraft and Burkhardt, 1986). (Mammalian rod terminals do not have telodendria, but they are contacted by telodendria from the cone terminals [Figure 3.9].) In some cases, however, there can be extensive interreceptor contacts above the level of the terminal. In toads, for example, radiating fins extend from the inner segments of the receptors, and extensive junctions occur between many of these processes (Custer, 1973; Gold and Dowling, 1979).

Most of the interreceptor contacts appear to be electrical junctions, and typical gap junctional morphology has been demonstrated at sites of contact between adjacent terminals, between telodendria, and between inner segment fins. But some interreceptor contacts may be chemical in nature. For example, two adjacent receptor terminals have been observed to contact each other adjacent to a synaptic ribbon (Lasansky, 1973), and telodendria from one terminal have been observed to penetrate into invaginations of adjacent cone terminals and make contacts at both ribbon and basal junctions (Normann et al., 1984; Kolb and Jones, 1985).

For the most part, large gap junctions are observed only between receptors of the same type; that is, cone–cone junctions or rod–rod junctions are much more prominent and larger than rod–cone junctions (Raviola and Gilula, 1973; Gold and Dowling, 1979). Rod–cone gap junctions in both primates and toads are small, linear junctions made up of a single row of connecting particles (Figure 3.5). Rod–cone interaction apparently is mediated by these tiny junctions (Nelson, 1977), but it appears to be weak (Gold, 1979; Attwell et al., 1984).

3.9 Electron micrographs showing inter-receptor contacts (*arrows*) made by primate receptor terminals (**a**) and by fins extending out from the inner segments of toad rods (**b**). The insets show higher power magnifications of a cone–rod junction (**a**) and a rod–rod junction (**b**). The cone–rod interreceptor contact is a tiny (kissing) gap junction (*wide arrows*); the rod–rod junctions are larger gap junctions. **a** from Dowling and Boycott (1966), reprinted with permission of the Royal Society. **b** modified from Gold and Dowling (1979), with permission of the American Physiological Society.

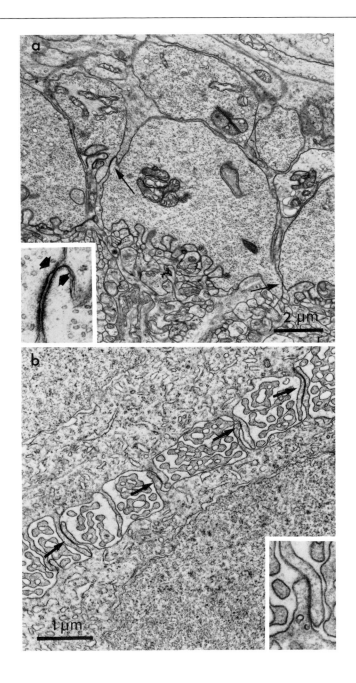

3.10 Electron micrographs of conventional synapses (*open arrows*) made by horizontal cell processes (H) in the mudpuppy retina. In both cases the synapses made by the horizontal cell processes appear to be onto bipolar cell dendrites (B). From Dowling and Werblin (1969), reprinted with permission of the American Physiological Society.

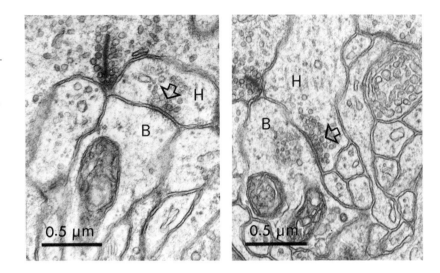

The role of the interreceptor junctions, especially the presumed chemical interactions between receptors, is not well understood. Surprisingly, even the extensive electrical interaction occurring between many rod cells does not appear to smudge the resolution of the receptor mosaic very much (Schwartz, 1976), and electrical coupling between photoreceptors of the same type may relate primarily to the physiology of the photoreceptor response (see Chapter 4).

HORIZONTAL CELLS   Chemical synapses made by horizontal cells are not observed frequently in mammals (but see Dowling et al., 1966; Fisher and Boycott, 1974). Many cold-blooded vertebrates, however, including amphibians (Dowling, 1968; Dowling and Werblin, 1969; Lasansky, 1973), reptiles (Kolb and Jones, 1984), and fishes (Sakai and Naka, 1986; Marshak and Dowling, 1987), show conventional synapses made by horizontal cells; in these animals the junctions are made mainly onto bipolar cell dendrites (Figure 3.10). Some synapses are close to the synaptic ribbon complex of the receptor terminal, but others are some distance away from the receptor terminals. Because horizontal cell synapses are not always observed readily, it has been suggested that horizontal cell synapses do not always have a conventional synaptic morphology and that, as I have already noted, the unusual invaginated synaptic complexes in the receptor terminal could allow for interac-

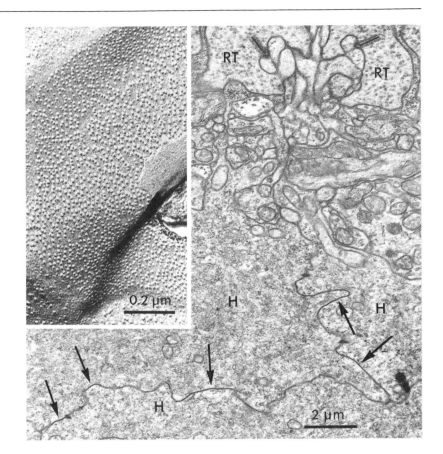

3.11 Electron micrograph showing extensive gap junctions (*arrows*) between adjacent horizontal cells (H) in the skate retina. The inset is from a freeze-fracture preparation showing the intramembranous particles at such junctions. The particles are much less tightly packed than at most non-horizontal cell gap junctions (see Figure 3.5). RT, receptor terminal. From Fain et al. (1976), reprinted with permission of Cold Spring Harbor Laboratory.

tions between horizontal, bipolar, and receptor cells in some way not understood. For example, the horizontal–receptor feedback synapse, which has been demonstrated physiologically in many retinas, has never been demonstrated anatomically. And even freeze-fracture analysis has not shown any intramembranous particle specializations on horizontal or receptor cell membranes to suggest synaptic feedback from horizontal cells to receptor cells (Schaeffer et al., 1982). In the catfish some conventional synaptic junctions between horizontal cell perikarya and receptor cell telodendria have been seen, but these junctions appear unique to this teleost (Sakai and Naka, 1986).

Much more prominent are the electrical (gap) junctions made by the horizontal cells (Figure 3.11). In fishes especially, these junctions are very extensive (Yamada and Ishikawa, 1965). But gap junctions between horizontal cell perikarya or between horizontal cell processes have been seen in many species (Kolb, 1977; Witkovsky et al., 1983); in almost all cases only horizontal cells of homologous types are coupled. In fishes, where this has been looked at especially carefully, H1 horizontal cells couple only with other H1 cells, H2 cells with other H2 cells, and so on (Witkovsky and Dowling, 1969; Kaneko and Stuart, 1984). In fishes and turtles the horizontal cell axon terminals are electrically coupled together; however, the axon terminals couple only with other axon terminals and do not form gap junctions with the cell perikarya of their own subtype of cells.

There seems to be little doubt that the effect of the gap junctions between the horizontal cells is primarily to increase the size of the receptive field of these cells (Plate 3). In fishes the dendritic diameters of the horizontal cell perikarya typically range from 30 to 150 μm (Stell and Lightfoot, 1975; Hassin, 1979), whereas the receptive fields of horizontal cells typically range from 2 to 5 mm in diameter (Naka and Rushton, 1967; Dowling and Ripps, 1971).

INTERPLEXIFORM CELLS   The synaptic organization of the interplexiform cells has been studied in some detail in teleost fishes, New World monkeys, cats, and humans (Dowling and Ehinger, 1975; Dowling et al., 1980; Kolb and West, 1977; Frederick et al., 1982). In all cases it has been shown that all of the input to these cells occurs in the inner plexiform layer, whereas most of the synapses made by interplexiform cells are found in the outer plexiform layer. Thus, these neurons appear to be primarily centrifugal in nature, carrying information from the inner to the outer plexiform layer.

In fishes and New World monkeys the interplexiform cells contain dopamine and can be marked selectively with certain monoamine analogues. When injected into the eye, the analogues are taken up by the dopaminergic cells and induce characteristic changes within the cells and their terminals (Figure 3.12). If the dose injected is not too high, the changes may be limited to the appearance of dense material within the synaptic vesicles. Such preparations can be used to map unequivocally the synapses made by these cells in both plexiform layers.

The interplexiform cells make conventional synapses. In the outer plexiform layer these have been observed onto both horizontal and bipolar cells in fishes and New World monkeys, onto bipolar cells and other interplexiform cell processes in the cat, and onto horizontal cells in the human retina. In the inner plexiform layer these neurons receive input mainly from amacrine cell pro-

3.12 Electron micrograph of the outer plexiform layer of a goldfish retina. Three dopaminergic terminals (*stars*) surround a horizontal cell (H). The retina was treated with a monoamine analogue, 5,6-dihydroxytryptamine, which causes electron-dense material to accumulate in the synaptic vesicles (*arrow and inset*). B, bipolar dendrite. From Dowling and Ehinger (1975 [inset], 1978a), reprinted with permission of the Royal Society and the AAAS.

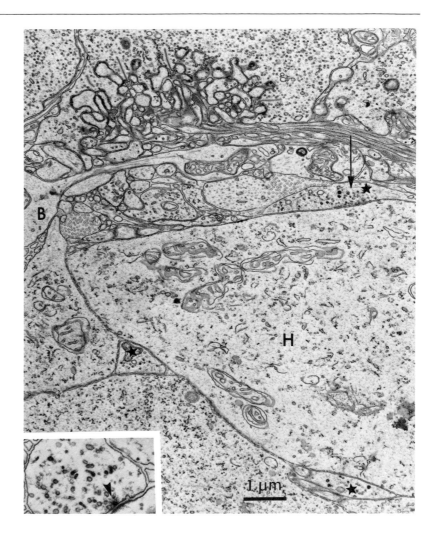

cesses, making only occasional synapses onto amacrine cell processes and perhaps (in the cat) onto bipolar cell processes. Figure 3.13 is a summary diagram of the synaptic connections of the interplexiform cells in the goldfish retina.

Inner Plexiform Layer

In all species the inner plexiform layer occupies a much greater volume than does the outer plexiform layer. Many more synapses, especially chemical synapses, are seen in the inner plexiform layer, and the synaptic structures appear more conventional than do

3.13 Schematic diagram of the synaptic connections of the interplexiform cells (IP) of the goldfish retina.

The input to these neurons is in the inner plexiform layer from amacrine cells (A). The interplexiform cell processes make synapses onto amacrine cell processes in the inner plexiform layer, but they never contact the ganglion cells (G) or their dendrites. In the outer plexiform layer the processes of the interplexiform cells surround the external (cone) horizontal cells (EH). They make synapses onto the external horizontal cell perikarya and onto bipolar cell dendrites. The interplexiform cell processes have never been observed as postsynaptic elements in the outer plexiform layer at either rod (R) or cone (C) receptor terminals or at the occasional external horizontal cell synapse. Nor are synapses seen between interplexiform cell processes and elements of the intermediate and internal horizontal cell layers.

IH, intermediate (rod) horizontal cell; EHA, external horizontal cell axon process; B, bipolar cell. From Dowling and Ehinger (1978a), reprinted with permission of the Royal Society.

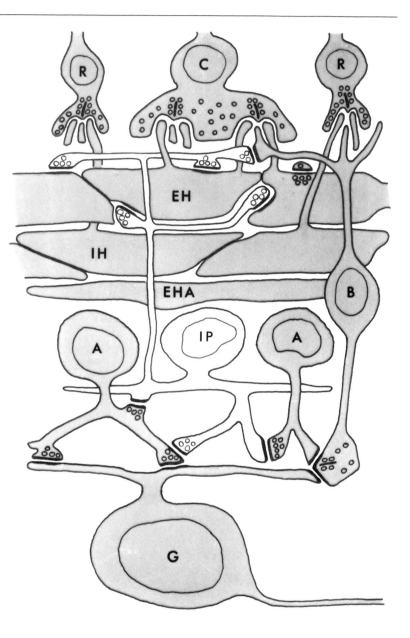

those of the outer plexiform layer (Kidd, 1962; Dowling and Boycott, 1966). Virtually all the synapses observed in the inner plexiform layer are made by either the bipolar terminals or the processes of amacrine cells. The interplexiform cells, the synaptic organiza-

3.14 Ribbon synapses (*filled arrows*) of bipolar terminals (B). Two postsynaptic processes (a dyad) are contacted at these junctions. One of the postsynaptic processes is an amacrine cell process (A); and that amacrine cell process makes a synapse (*open arrows*) onto another process (*left*) or back onto the bipolar cell terminals (*right*). Micrograph on the left is from a chicken; on the right from a skate. From Dowling (1979), reprinted with permission of MIT Press.

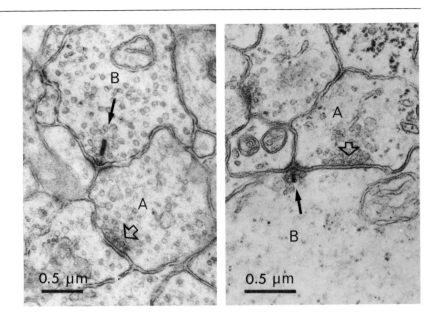

tion of which was discussed above, and centrifugal fibers, when present, make only a relatively few synapses in this layer.

BIPOLAR CELLS   Bipolar cells appear to make only ribbon synapses. Although occasionally it has been suggested that bipolar terminals make conventional synapses (Allen, 1969), these reports have not been substantiated. The ribbons in the bipolar terminals, as already noted, tend to be much smaller than those seen in the receptor terminals. As in the photoreceptor terminals, however, the ribbon is oriented perpendicular to the terminal membrane and sits above an evaginated ridge along the terminal membrane (Figure 3.14). An arciform density is observed between ribbon and membrane, and the ribbon is surrounded by a halo of synaptic vesicles. The ribbon synapses of the bipolar cells are never invaginated like those of the receptor terminals, even though in some species the bipolar terminals, especially rod bipolar terminals, are as large as many of the receptor terminals.

In all species, almost invariably there are two postsynaptic processes apposed to the bipolar terminal synaptic ribbons, and this postsynaptic arrangement has been called a dyad (Dowling and

Boycott, 1966). On occasion, three processes may be observed at a ribbon bipolar synapse or, very rarely, just one process (Dubin, 1970). Typically, prominent electron-dense material is associated with the postsynaptic processes at bipolar ribbon synapses, providing a distinct marker for those elements associated with the complex.

The two postsynaptic processes of a dyad are most often a ganglion cell dendrite and an amacrine cell process or two amacrine cell processes (Dowling, 1968). In some retinas the former situation appears to predominate, whereas in other retinas the reverse is true (see below). Occasionally two ganglion cell dendrites make up a dyad, but this is very rare indeed. No substantial evidence of bipolar cells synapsing on other bipolar terminals at ribbon synapses has been presented.

There have been reports of gap junctions between bipolar cells in some species (Witkovsky and Stell, 1973; Kujiraoka and Saito, 1986) and of gap junctions between bipolar terminals and other elements in the inner plexiform layer of some retinas. The best-analyzed instance of this latter arrangement involves the rod pathway found in the cat (see below).

AMACRINE CELLS Numerous conventional synapses made by amacrine cell processes are seen in the inner plexiform layer of all species. Such amacrine cell synapses are made onto ganglion cell dendrites, bipolar cell terminals, interplexiform cell processes, and other amacrine cell processes (Dowling and Boycott, 1966). Often a synapse made by an amacrine cell process is seen just adjacent to a site where the amacrine cell process itself is postsynaptic to a ribbon or conventional synapse. Thus, amacrine cell processes can be both pre- and postsynaptic over a very short portion of their length (Figures 3.14 and 3.15). Furthermore, as noted earlier, electron microscopy has shown that amacrine cell processes have characteristics of both axons and dendrites and, for the most part, no segregation of pre- and postsynaptic membranes has been observed (but see below). That is, amacrine cell processes can both make and receive synapses all along their length.

Because amacrine cell processes can be both pre- and postsynaptic over short distances, they make two types of synaptic arrangements of note. First, at many ribbon synapses of the bipolar terminals, a postsynaptic amacrine cell process makes a synapse back onto the bipolar terminal a short distance away (Figure 3.14). The synapses involved in this arrangement are called reciprocal synapses; and this configuration suggests that a local, feedback interaction could occur between bipolar terminals and amacrine processes at the ribbon synapses. Reciprocal synapses involving just amacrine processes have also been seen (Figure 3.15).

The synapses involved in the second arrangement are called serial synapses, and this configuration suggests the possibility of local, serial interactions between adjacent amacrine cell processes (Kidd, 1962; Dowling, 1968). In such arrangements, one amacrine cell process makes a synapse onto an adjoining amacrine cell process, which in turn makes a nearby synapse onto a third element. The third element may be a ganglion cell dendrite, a bipolar cell terminal, or another amacrine cell process. In some species more abundant amacrine cell interaction is observed, and the serial synapsing may be more extensive and involve as many as four consecutive synapses (Figure 3.15).

Gap junctions between amacrine cell processes are occasionally seen (Figure 3.5); and, although they have not been studied in detail, the prevailing view is that these junctions, like those between horizontal cells and their processes, serve to extend the receptive field size of the cells (Naka and Christensen, 1981). Because ama-

3.15 Serial and reciprocal synapses (*arrows*) made by amacrine cell processes in the inner plexiform layer of the frog. This micrograph illustrates clearly that amacrine cell (A) processes may be both pre- and postsynaptic. From Dowling (1968), reprinted with permission of the Royal Society.

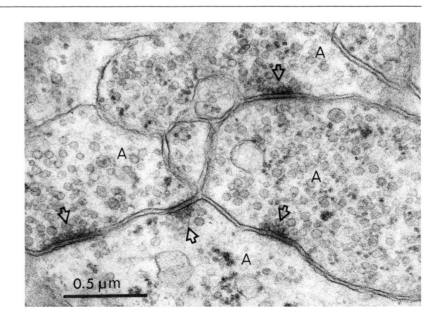

crine cell size varies so greatly in most retinas, it seems unlikely that all amacrine cell types are connected by gap junctions.

A prominent amacrine process–bipolar terminal gap junction in the cat retina appears to involve the rod pathway in that animal (Famiglietti and Kolb, 1975), and this finding and several associated observations, although limited to a single species, are of such interest as to warrant special discussion (Kolb, 1979). Rod bipolar terminals in cats extend deep into the inner plexiform layer, where they make ribbon synapses only with amacrine cell processes. At each ribbon synapse, processes from two types of amacrine cells are usually contacted. One type, called the AI amacrine cell, usually makes a feedback synapse onto the bipolar terminal; the other type does not. This other type, called the AII amacrine cell, makes no synapses at this level of the inner plexiform layer; but somewhat higher in the layer it makes extensive gap junctions with invaginating cone bipolars. Still higher in the inner plexiform layer the AII amacrine makes conventional synapses onto ganglion cell dendrites, flat bipolar terminals, and other amacrine cell processes. It also receives some conventional synaptic input from other amacrine cells.

Two interesting and important conclusions can be drawn from these observations. First, rod bipolars in the cat retina do not directly contact ganglion cell dendrites; rather all signals from rod bipolar cells go through amacrine cells before they reach the ganglion cells (see Nelson, 1982). Thus, rod information traveling this pathway must pass across at least three synapses (rod to bipolar to amacrine to ganglion cell) before leaving the eye, and rod information reaching the ganglion cells via the invaginating cone bipolars passes through at least four synapses, including an electrical junction, before reaching the optic nerve.* Second, the AII amacrine cell appears to have its synaptic interactions segregated to some extent. It receives much, if not most, of its input deep in the inner plexiform layer from ribbon synapses of rod bipolars, it makes gap junctions with invaginating bipolars in the middle of the layer, and it makes conventional synaptic contacts with bipolar, ganglion, and amacrine cells in the upper part of the inner plexiform layer.

---

* Rod signals in the cat can also spread to cone receptor terminals via interreceptor contacts and pass on to ganglion cells by that route (Nelson, 1977; Smith et al., 1986). It has been proposed that the pathway rod signals traverse to reach the ganglion cells may depend on the intensity of the light stimulus (Smith et al., 1986).

CENTRIFUGAL FIBERS   Centrifugal fibers have been identified
with certainty in the retina of several cold-blooded vertebrates and
birds. Although reports going back at least as far as Cajal (Ramón
y Cajal, 1911) have indicated centrifugal fibers entering the mammalian retina from the optic nerve, these findings have not been
substantiated; and the common view is that there are very few if
any such fibers in the mammal (but see Honrubia and Elliott,
1968).

The centrifugal fiber system of the bird has been studied in some
detail (Figure 3.16). The nucleus of origin of the fibers in the mid-
brain, the isthmo-optic nucleus, was identified (Cowan and Powell,
1963) and the synapses of the fibers were located by injuring the
cells in this nucleus and searching for degenerating fibers in the
retina. The centrifugal fiber terminals make conventional synapses
mainly along the proximal margin of the inner plexiform layer, on
the basal portions of amacrine cells (Dowling and Cowan, 1966).
Occasionally centrifugal fiber synapses are seen deeper in the inner
nuclear layer, on the cell perikarya of larger neurons. These larger
postsynaptic cells may be a separate type of amacrine cells (Cajal's
association amacrine of the bird) or possibly displaced ganglion
cells (Maturana and Frenk, 1965). In a teleost fish (white perch),

3.16 Drawing of centrifugal fibers
in the bird retina and their termina-
tions along the innermost aspect of
the inner nuclear layer (based on
Golgi-stained material). From Dowling and Cowan (1966), redrawn from
Cajal; reprinted with permission of
Springer-Verlag.

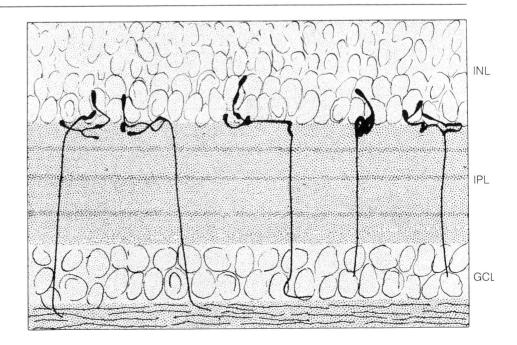

INL

IPL

GCL

at least some centrifugal fibers synapse onto the perikarya and proximal dendrites of interplexiform cells (C. L. Zucker and J. E. Dowling, unpublished observations).

Comparative Aspects of Synaptic Organization

When the patterns of synaptic organization of different species are compared, some striking and significant differences are apparent, particularly with regard to the inner plexiform layer (Dowling, 1968; Dubin, 1970). There are almost certainly some differences between species in the synaptic organization of the outer plexiform layer also, but these are not as obvious as those of the inner plexiform layer. Furthermore, physiological studies also suggest that there is more variability between species in the inner plexiform layer than in the outer plexiform layer.

In the central part of the inner plexiform layer of primates, the postsynaptic elements observed at the bipolar ribbon synapses consist, in about 80 percent of the cases, of a ganglion cell dendrite and an amacrine cell process. Similarly, at about 90 percent of the ribbon synapses of cone bipolars in the cat, at least one member of the dyad is a ganglion cell dendrite (Kolb, 1979). Thus, much of the cone bipolar output in both primates and cats is directly onto ganglion dendrites. In frogs, on the other hand, at over 70 percent of the ribbon synaptic contacts in the inner plexiform layer, both postsynaptic elements appeared to be amacrine cell processes. Thus, in frogs, most of the bipolar cell input to the inner plexiform layer appears to be onto amacrine cell processes (Dowling, 1968).

There are also striking differences in the number of amacrine cell synapses seen in the inner plexiform layer of different species. In central primate retina, conventional amacrine cell synapses were observed at a frequency of $0.047/\mu m^2$, whereas in frogs, conventional synapses in the inner plexiform layer were observed with a frequency of $0.212/\mu m^2$ (Dowling, 1968), over a fourfold difference. Furthermore, serial synapses involving conventional synapses are relatively rare in primates but are observed frequently in frog retinas.

The incidence of bipolar ribbon synapses per unit area was found to be similar in primates and frogs (0.028 and $0.024/\mu m^2$), and this is also true for other species (Kidd, 1962; Dubin, 1970). Thus, a convenient way to quantify differences in the relative frequency of inner plexiform layer synapses between species is to determine the ratio of conventional to ribbon junctions. This measurement was made by Dubin (1970) for eight species, using similar preparation methods and criteria for judging synaptic contacts; his results are shown in Table 3.1. Four groupings appear: the cone-rich central region of the primate retina has the lowest amacrine to bipolar cell synapse ratio; whereas the peripheral retina of monkeys and the retinas of cats and rats have a ratio about 50 percent higher. Rabbits and ground squirrels have ratios about three times that of the central primate retina; and frog and pigeon retinas show the highest ratios, about five times that found in the central primate retina.

Put another way, over 90 percent of the synapses observed in the inner plexiform layer of the frog or pigeon are amacrine cell synapses; but in central primate retina at least 35 percent of the synapses observed are bipolar synapses. Furthermore, because the number of ribbon synapses per unit area is approximately the same in primates and frogs, the total number of synapses observed per unit area is much higher in the frog than in primates (10,400,000 contacts/mm^2 and 2,900,000 contacts/mm^2, respectively). These data indicate that there is much more amacrine cell synaptic interaction in some retinas than in others and suggest that amacrine cells are likely to have a much more important role in the processing of visual information in certain animals.

A third difference in synaptic organization between species is found in the input to the ganglion cells. As already noted, ribbon synaptic input to the ganglion cell dendrites is abundant in primates and rare in frogs. But many conventional synapses are observed on all frog ganglion cell dendrites. These qualitative observations suggest that certain ganglion cells receive predominantly amacrine cell

Table 3.1 Inner plexiform layer synapses

Species	Ratio of conventional to ribbon synapses	Percentage of synapses in serial configuration
Human (central)	1.7:1	1.9
Monkey (central)	1.9:1	7.1
Monkey (periphery)	3.0:1	5.1
Cat	2.8:1	7.8
Rat	3.3:1	6.7
Rabbit	5.2:1	12.6
Ground squirrel	6.7:1	10.1
Frog	9.6:1	13.4
Pigeon	10.8:1	11.2

input, whereas other ganglion cells receive abundant bipolar cell input. When considered in conjunction with the data of Table 3.1, these observations suggest that ganglion cells that receive abundant bipolar cell input predominate in some species (primates, for example), whereas ganglion cells that receive mainly amacrine cell input predominate in others (frogs and pigeons, for example). The obvious question is whether those species that are intermediate between primates and frogs or pigeons in terms of their ratios of inner plexiform layer synapses (rabbits and ground squirrels) have mainly ganglion cells with a mix of inputs from amacrine and bipolar cells or separate populations of ganglion cells, some of which have predominately amacrine cell input and others that have abundant bipolar input.

Evidence in favor of the latter suggestion was provided by a serial section study of Golgi-stained ganglion cells of the ground squirrel retina (West and Dowling, 1972; West, 1976). In that study fifteen different types of ganglion cells were distinguished on the basis of differences observed in the light microscope. In terms of synaptic input, however, the ganglion cells could be separated into two groups. One group (called group A) had a very high percentage of amacrine cell input (averaging 93 percent, with a range of 78 to 100 percent) and the other group (called group B) had a predominance of bipolar cell input. The average percentage of bipolar cell input into the group B cells was 59 percent, with a range of 46 to 79 percent. Table 3.2 gives the numerical data obtained in this study, and several of the ganglion cells studied are shown in Figure 2.10. Note that at least one type (1a) of the group A cells (and perhaps two) had no bipolar input at all. Three cells of type 1a were extensively examined (one in its entirety), and no ribbon synapses onto their dendrites were observed, although nearly 100 conventional synaptic inputs onto one cell were seen. This finding is important because it shows unequivocally that some ganglion cells may be at least fourth-order neurons along the visual pathway.

A more recent study in the cat retina has confirmed these general ideas (Kolb, 1979). Ganglion cells believed to be sustained, contrast-sensitive (X) cells, which are the predominate ganglion cells in the cat retina and account for about 55 percent of the cell population, were found to have approximately 70 percent bipolar input and 30 percent amacrine input. Other ganglion cells, however, were found to have 75 to 80 percent amacrine cell input, a finding indicating that, in this animal also, certain ganglion cells have

Table 3.2 Synaptic input to ganglion cell types in ground squirrel retina

Type of cells	Number of cells examined	Input (%) B	Input (%) A	Synapses observed	Dendritic spread (μm)
Group A					
1a	3	0	100	207	50
2	2	4	96	156	60
3	1	2	98	49	60
4	1	0	100	9	50
5	1	9	91	44	70
6	2	3	97	106	100
7	2	3	97	29	150
8	1	7	93	55	90
10	2	19	81	79	100
12	1	22	78	23	100
14	1	12	88	25	>350
Total	17	7	93	782	
Group B					
1b	2	76	24	46	50
9	3	46	54	123	30
11	2	47	53	40	150
13	3	48	52	27	500
15	1	79	21	42	200
Total	11	59	41	278	

abundant bipolar cell input whereas others are driven primarily by the amacrine cells.

Physiological Correlations

Can one suggest correlations between the synaptic organization of the retina and its physiological responses? Consideration of the ground squirrel retina is again instructive. This retina has virtually all cones, with no more than 5 percent rodlike photoreceptors (West and Dowling, 1975). It has both major functional classes of ganglion cells described in Chapter 1 (Michael, 1968a,b,c). That is, one class of ganglion cells in the ground squirrel responds in a sustained fashion to spot or annular illumination, and their receptive fields can be mapped into center and surround antagonistic regions (Figure 2.13). These are typical contrast-sensitive cells. (Some cells of this class in the ground squirrel show color-opponent characteristics; that is, the center and surround regions are each driven by cones of a different spectral class and the antagonistic regions of these fields may overlap partially or completely.) Cells of the second major class of ganglion cells in the ground squirrel respond with transient responses to spots of light presented to the centers of their receptive fields; these cells give bursts of impulses at the onset and offset of illumination and are always movement- and direction-sensitive. Their fields closely resemble the model receptive field shown in Figure 2.14.

There is a further difference in the receptive fields of these two classes of ganglion cells in the ground squirrel: the contrast-sensitive cells tend to have a larger center field than the direction-sensitive cells. This size difference provides a convenient means of correlating the anatomy with the physiology. All group A cells, except for one type (Table 3.2), have small dendritic fields (40–100 μm), whereas a number of the cells in group B have much larger dendritic spreads, 150 to 500 μm. Michael found that the direction-sensitive cells had small receptive field centers (60–115 μm) and that the contrast-sensitive cells had receptive field centers ranging from 60 to 500 μm. These data suggest that the on–off, direction-sensitive cells receive most, if not all, of their input from amacrine cells and that the contrast-sensitive cells receive about half of their input from bipolar cells. These data also suggest that complex processing such as movement and direction selectivity occurs in the inner plexiform layer and is mediated by amacrine cells.

These ideas are supported by evidence from other species. Both frogs and pigeons are known to possess many transient ganglion cells that are movement- and direction-sensitive and that show other complex response properties (Barlow, 1953; Maturana et al., 1960; Maturana and Frenk, 1963). The complexity of ganglion cell responses in these animals corresponds well with the large number of amacrine cell synapses seen in their inner plexiform layers (Table 3.1). Conversely, many cells in the cat retina and in the central part of the primate retina respond to spots of light in a sustained fashion and show a center–surround receptive field organization (Hubel and Wiesel, 1960; Gouras, 1968). In the cat, as already noted, such cells (the X-cells) receive about 70 percent of their synaptic input from bipolar cells (Kolb, 1979), and the low ratio of amacrine to bipolar synapses found in the central region of the primate retina suggests that there also the ganglion cells receive mainly bipolar input. Thus, it would appear that amacrine cell interactions are less important in the generation of contrast-sensitive receptive fields. These notions receive powerful support from both physiological and pharmacological evidence (Chapters 4 and 5).

Scheme of the Synaptic Organization of the Retina

Figure 3.17 is a schematic diagram of the synaptic contacts observed in many retinas and serves as a summary of the synaptic organization of the vertebrate retina. It pertains mainly to the cone pathways in the retina; less is known about rod pathways in most species. Rod pathways in the cat have already been discussed, and it may be that similar rod pathways exist in other mammals (Dacheux and Raviola, 1986). In nonmammalian species, rod and cone pathways through the retina are presumed to be similar, because, unlike the bipolar cells in mammals, many bipolar cells in these species receive direct input from both rods and cones (Chapter 2). For simplicity, only the chemical synapses are illustrated. Although this drawing was originally prepared to illustrate synaptic contacts made in one species (frog), it is now clear that the synaptic pathways illustrated are rather general. Differences between species are more often quantitative rather than qualitative; that is, all of the synaptic arrangements shown exist in virtually all retinas, although certain retinas have more of one kind of arrangement than of another. This is especially true of the inner plexiform layer.

In the outer plexiform layer (upper half of Figure 3.17), the receptor terminals make junctions with the processes of horizontal cells and the dendrites of the bipolar cells. The flat bipolar cells are postsynaptic at basal junctions and the invaginating bipolar cells postsynaptic at ribbon synapses (Kolb, 1970). Freeze-fracture evidence suggests that the junction between a receptor and a flat bipolar cell has characteristics of an excitatory synapse, whereas the junction between a receptor and an invaginating bipolar cell resembles inhibitory synapses seen elsewhere in the brain (Raviola and Gilula, 1975; Schaeffer et al., 1982).

As stated above, horizontal cells contact the receptor terminals at the ribbon synapses. These contacts usually occur within invaginations of the receptor terminal, and often there is a precise arrangement of processes within the invagination (Stell, 1965; Missotten, 1965). Typically two horizontal cell processes penetrate deeply into the invagination and come to lie on either side of the ribbon. The more superficially and centrally positioned processes in the invagination are usually the dendrites of the bipolar cells. What the significance of the invagination is, or what the precise arrangement of the processes within the invagination means, is still obscure. As already noted, one idea is that the invagination facili-

3.17 Summary diagram of the arrangements of synaptic contacts found in vertebrate retinas.

In the outer plexiform layer, processes from invaginating bipolar (IB) and horizontal (H) cells penetrate into invaginations in the receptor terminals (RT) and terminate near the synaptic ribbons of the receptors. The processes of flat bipolar cells (FB) make superficial (basal) contacts onto the bases of some receptor terminals. Horizontal cells make conventional synaptic contacts onto bipolar dendrites. The processes of the interplexiform cells (IP) make conventional synapses onto horizontal cells or bipolar cells or both.

In the inner plexiform layer, bipolar terminals most commonly contact one ganglion cell (G) dendrite and one amacrine cell (A) process at the ribbon synapse (*left side of drawing*), or two amacrine cell processes (*right side of drawing*). When the latter arrangement predominates in a retina, numerous conventional synapses between amacrine cell processes (serial synapses) are observed. Amacrine cell synapses in all retinas make synapses back onto bipolar cell terminals (reciprocal synapses).

The input to ganglion cells may differ in terms of the proportion of bipolar and amacrine synapses. Ganglion cells may receive mainly bipolar cell input (G_1), a more even mixture of bipolar and amacrine cell input (G_2), or mainly or exclusively amacrine cell input (G_3). The interplexiform cells receive input from amacrine cells, and they occasionally make synapses on other amacrine cell processes. Modified from Dowling (1968), with permission of the Royal Society.

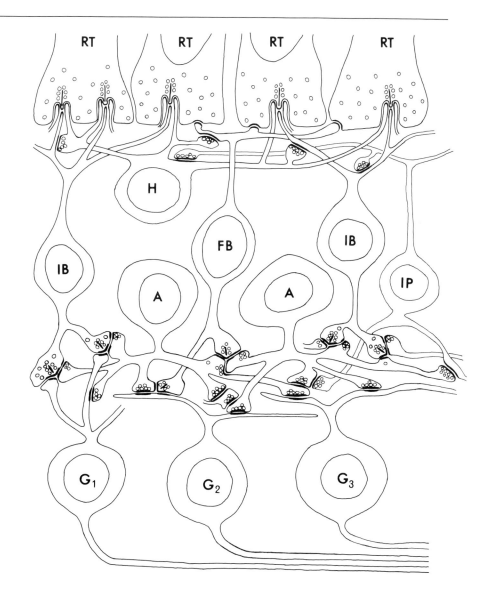

tates interactions between the horizontal cell processes, the bipolar processes and the receptor terminal (Dowling and Boycott, 1966; Raviola and Gilula, 1973).

In many, but not all, species, synapses made by the horizontal cells have been identified. These synapses are of the conventional type and are made predominantly onto bipolar dendrites. Horizontal cell synapses feeding back onto receptor terminals have been seen only very rarely, even though physiological evidence for feedback, especially onto cones, is strong. Thus, the question remains as to how horizontal cells synaptically interact with both bipolar cells and receptor terminals.

The principal synaptic interactions occurring in the outer plexiform layer can be summarized as follows: receptors clearly make chemical synapses with both horizontal and bipolar cells. Horizontal cells appear to interact with both bipolar cell dendrites and receptor terminals, but ultrastructural correlates of these synapses have not been firmly identified in many cases. Because horizontal cells spread their processes much further laterally in the outer plexiform layer than do bipolar cells, the cellular and synaptic organization of this layer suggests that bipolar cells are driven directly by nearby receptors and indirectly by distant receptors via the horizontal cells.

The remaining chemical synapses observed in the outer plexiform layer are those of the interplexiform cells. In fishes and New World monkeys interplexiform cells make synapses onto both horizontal and bipolar cells and their processes; in cats such synapses have been identified only onto the bipolar cells and their dendrites; in humans interplexiform cells synapse onto horizontal cells.

The inner plexiform layer of the retina (lower half of Figure 3.17) is thicker than the outer plexiform layer in all vertebrates, more synaptic contacts per unit area are seen, and a greater variety of junctions is observed. Bipolar cell terminals contact amacrine cell processes and ganglion cell dendrites at ribbon synapses. In virtually all cases, two postsynaptic elements are observed at these synapses and are called a dyad. The postsynaptic elements making up the dyad may consist of a ganglion cell dendrite and an amacrine cell process or two amacrine cell processes, or, very rarely, two ganglion cell dendrites. Which of the pairings predominates depends on species and type of bipolar cells.

In all retinas numerous conventional synapses made by amacrine cell processes are observed in the inner plexiform layer. Amacrine cells make synapses onto ganglion cell dendrites, onto bipolar cell terminals, and onto other amacrine cell processes. The synapses involving amacrine cell processes may be organized in a serial or reciprocal fashion. Such synaptic arrangements suggest the possibility of local interactions between amacrine cell processes and other elements within the inner plexiform layer. Finally, occasional conventional synapses received by or made onto and between the interplexiform cell processes are observed in the inner plexiform layer. The synapses involving interplexiform cell processes are mainly with amacrine cell processes.

Evidence for alternative synaptic pathways in the inner plexiform layer have come from comparative studies. When the anatomies of various retinas are compared, we find that retinas that have many contrast-sensitive ganglion cells (primates and cats) have many dyad pairings consisting of one amacrine cell process and one ganglion cell dendrite, a relatively low number of amacrine cell synapses per unit area, and few serial synapses. Retinas with abundant ganglion cells showing complex receptive field properties such as movement or direction selectivity (frogs and pigeons) have dyad pairings consisting mainly of two amacrine cell processes, abundant amacrine cell synapses per unit area, and many serial synapses. Retinas with numerous examples of both kinds of ganglion cells (ground squirrels and rabbits) are intermediate in terms of numbers of amacrine cell synapses per unit area.

These observations indicate that (1) in retinas where the simple, sustained, contrast-sensitive type of receptive field predominates, bipolar terminals make numerous direct contacts with ganglion cell dendrites, whereas in retinas where complex, motion- and direction-sensitive types of receptive fields predominate, relatively fewer direct bipolar–ganglion cell contacts occur; and (2) there are significantly more amacrine synapses and amacrine–amacrine interactions in retinas with complex receptive fields than in retinas with simpler receptive fields. These ideas are incorporated in the lower half of Figure 3.17. The left side of the drawing represents a "simple" inner plexiform layer organization: the ganglion cell (G_1) receives its input mainly via direct bipolar–ganglion cell junctions, and there are relatively few amacrine cell junctions. Such a ganglion cell presumably would have a contrast-sensitive receptive field similar to that shown in Figure 2.13. On the right side, a "complex" inner plexiform layer organization is pictured: the ganglion cell (G_3) receives its input mainly or exclusively from amacrine cell processes, which make many contacts among themselves and onto the ganglion cell dendrites. This ganglion cell presumably would show complex receptive field properties, perhaps similar to those of Figure 2.14. Finally, the ganglion cell in the middle (G_2) is shown as receiving a more equal input from the bipolar and amacrine cells. Presumably this cell would show a mix of receptive field properties; such ganglion cells are known (see Chapter 2).

A Final Comment: Morphological Diversity of Ganglion Cells
I have implied in Figure 3.17 and the related discussion that the
receptive field properties of a ganglion cell depend mainly on the
kind of input the cell receives, whether it is predominantly from
bipolar or amacrine cells. There is much physiological evidence to
support this view (Chapter 4). What remains unclear is the significance of the variation in morphology of ganglion cells, especially
of their dendritic trees, and the relation of these variations to the
receptive field organization of these cells. Clearly a ganglion cell
whose dendrites are confined to the distal half of the inner plexiform layer, where the flat (off) bipolars mainly terminate, will receive a synaptic input different from that of a ganglion cell whose
dendrites run in the proximal part of the inner plexiform layer,
where the invaginating (on) bipolars mainly terminate. One obvious notion is that the level of dendritic branching in the inner
plexiform layer determines the connectivity onto the cell. Furthermore, as shown by the data of Table 3.2 on ground squirrel ganglion cells, different morphological types of ganglion cells have different ratios of synaptic input. This difference is consistent and
enables one to predict in most instances the input ratio of a ganglion cell on the basis of its morphological appearance. This is not
always the case, however. Ganglion cell types 1a and 1b in ground
squirrels appear morphologically indistinguishable (a and b in Figure 2.10); their dendrites are found at the same level in the inner
plexiform layer, and they appear to have a similar branching pattern and dendritic spread. But they have strikingly different synaptic inputs. Type 1a has 100 percent amacrine input, whereas type
1b has 78 percent bipolar input and 22 percent amacrine input.
Presumably these two morphologically indistinguishable cell types
have very different physiological responses.

It is also important to note that relatively few ganglion cell receptive field types have been described in ground squirrels, even
though there are many morphologically distinct ganglion cells
(Table 3.2). Many of the morphologically distinct ganglion cells,
however, have a very similar amacrine to bipolar cell ratio (that is,
types 2, 3, 5, 6, 7, and 8). It may be that many of these cells have
a similar basic receptive field organization and differ only in subtle
ways that have not been detected as yet physiologically. Also, additional receptive field types in the ground squirrel may each be
represented by only a small percentage of the ganglion cell population, as in the rabbit (Caldwell and Daw, 1978a), and these may

be those cell types with the somewhat different amacrine–bipolar input ratios (for example, types 10, 12, and 14). Clearly more information on the synaptic input and receptive field properties of identified cells is needed to provide insights into the functional significance of the morphological diversity of the ganglion cells and other retinal neurons.

Neuronal Responses

A MAJOR reason the retina has been a particularly fruitful region of the central nervous system to study is that intracellular recordings can be made from all classes of retinal cells. In most parts of the vertebrate brain virtually all the information we have concerning the electrical properties of the neurons has come from extracellular recordings from single cells or from populations of cells. Such extracellular recordings generally record only action potentials and thus provide information concerning the kind of information processing occurring in that part of the brain. But they provide little or no information concerning the mechanisms underlying the generation of the responses. And if there are neurons that do not generate action potentials, as is the case in the retina, extracellular recordings will usually fail to detect them.

In retinas of mammalian and avian species the retinal neurons are small, and intracellular recording is difficult. In cold-blooded vertebrates, however, the retinal neurons are usually considerably larger (compare Figures 2.1 and 2.2), and intracellular recordings have been made from most, if not all, of the types of retinal cells. Cold-blooded animals that have been particularly useful for intracellular recordings include the mudpuppy, goldfish–carp, turtle, catfish, and tiger salamander. More recently, a number of successful intracellular recordings from various retinal neurons have been made in both cat and rabbit retinas, so intracellular retinal responses from all major groups of animals except birds have now been reported. The results show that the response characteristics of retinal neurons are similar across species. Thus, from both intracellular and extracellular recording data, a number of generalizations can be made and inferences drawn concerning the functional organization of the vertebrate retina.

Intracellular Recordings

The first intracellular recordings from the vertebrate retina were made with glass micropipettes by Gunnar Svaetichin in 1953. He was working with the fish retina and believed he was recording from the photoreceptors. Later studies indicated that he almost certainly was recording from the large horizontal cells found there (MacNichol and Svaetichin, 1958).* As this example illustrates, a major problem in the early days of intracellular recordings from the retina was to identify the type of cell from which the recording was made. Initial intracellular staining studies used dyes such as Niagara Sky Blue, which provides a mark at the site of recording within the cell, usually in the cell perikaryon (Bortoff, 1964; Kaneko and Hashimoto, 1967; Werblin and Dowling, 1969). Because the perikarya of many retinal cells have distinctive shapes and are localized to specific layers within the retina, such experiments identified the recorded cell reasonably well. Subsequently, most intracellular staining in the retina has been done with dyes such as Procion Yellow or Lucifer Yellow. These dyes have the great advantage of diffusing throughout the cell, into both dendritic and axonal processes (Kaneko, 1970), and identification of retinal cell types after such staining is usually unequivocal (Plates 1, 2, and 3). In recent years retinal cells have been injected with horseradish peroxidase.This compound also diffuses throughout the cell; but, in addition, it can be made electron dense so that the stained cell can be visualized in both the light and electron microscopes (see, for example, Lasansky, 1978). In materials stained with horseradish peroxidase the synapses made onto the cell, and sometimes by the cell, can be studied (Plate 4).

An early and surprising finding derived from intracellular recordings from the retinal neurons was that the distal neurons that synapse in the outer plexiform layer—the receptors, the horizontal cells, and the bipolar cells—respond to retinal illumination only with sustained, graded potentials (Figure 4.1). Within the retina these cells ordinarily do not generate classic action potentials, although recent work, particularly on isolated and cultured retinal neurons, has shown that these neurons have a variety of voltage-sensitive channels (Werblin, 1975; Fain et al., 1977; Tachibana, 1983; Kaneko and Tachibana, 1985a) including voltage-sensitive Na^+ channels similar to those found in action potential-generating axons (Shingai and Christensen, 1983; Lasater, 1986). The neurons in the proximal retina—the amacrine and ganglion cells—are

---

* The graded potentials recorded by Svaetichin were originally called cone action potentials. When it became clear that these potentials arise more proximally in the retina, they were called S-potentials, in recognition of their slow, sustained nature and of their discoverer. The term S-potentials is now used exclusively for horizontal cell potentials.

Plate 1 Bipolar cell from the carp retina stained intracellularly with Procion Yellow. Micrograph provided by M. Tachibana and A. Kaneko.

25 μm

Plate 2 Starburst amacrine cell from the rabbit retina stained intracellularly with Lucifer Yellow and viewed in a flat mount preparation. From Tauchi and Masland (1984), reprinted with permission of the Royal Society.

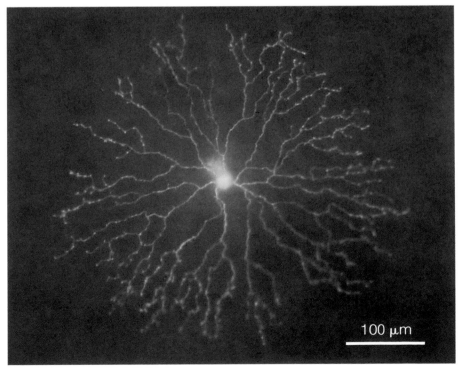

100 μm

Plate 3 Horizontal cell perikarya from the white perch retina stained intracellularly with Lucifer Yellow and viewed in flat mount preparations. The cell on the left is from a retina treated with dopamine, which decreases electrical coupling between horizontal cells and restricts the movement of dye from the injected cell. The micrograph on the right is from a control preparation in which dye spread from the injected cell (*center*) to many surrounding cells via electrical (gap) junctions. Micrographs provided by T. Tornqvist and X.-L. Yang.

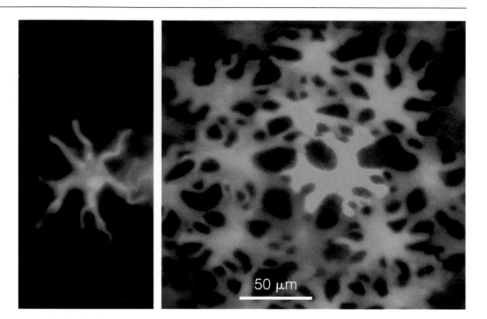

Plate 4 Ganglion cell from the catfish retina stained intracellularly with horseradish peroxidase (HRP). Inset is an electron micrograph showing a ribbon synapse (*arrow*) onto an HRP-stained ganglion cell dendrite. Micrographs provided by H. Sakai and K.-I. Naka.

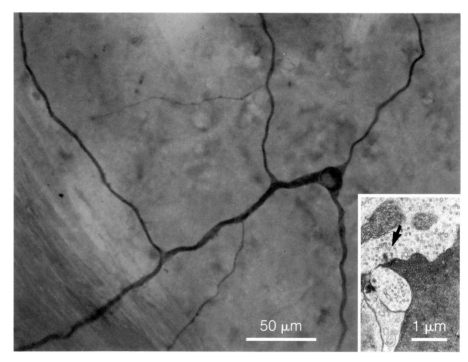

Plate 5 Monoamine-containing amacrine cells in the rabbit retina visualized by the Falk-Hillarp fluorescence method. The indoleamine-containing cells show a yellower fluorescence than do the catecholamine-containing cells (*inset*). Transparency provided by B. Ehinger.

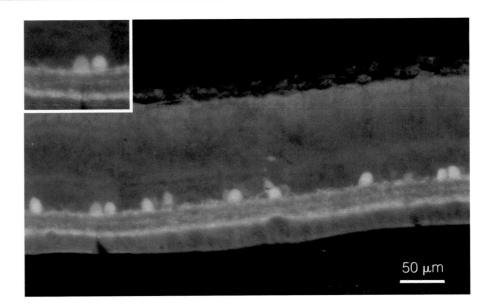

Plate 6 Dopamine-containing interplexiform cells in the cichlid fish retina visualized by the Falk-Hillarp method. The catecholamine fluorescence appears yellower than usual because of the filters used to take the photograph. Transparency provided by A. Laties.

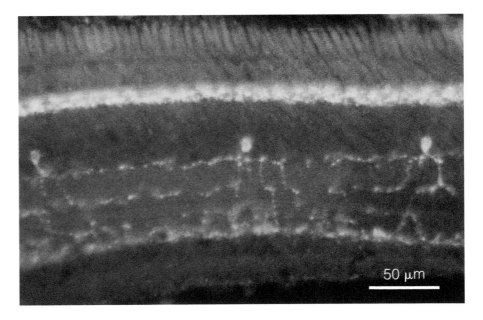

Plate 7 Bullfrog retinas (*top*) and extracts of bullfrog retinas (*bottom*) showing the color of the rod visual pigment, rhodopsin (*left*). Following exposure to light, rhodopsin bleaches first to retinal (vitamin A aldehyde) and protein (opsin) (*middle*), and finally to vitamin A and opsin (*right*). Retinas and extracts prepared by P. K. Brown and J. E. Dowling.

Plate 8 Difference spectra (the difference in light absorption before and after bleaching) of the three human and monkey cone pigments and human rhodopsin. The colored symbols above the spectra (cone pigments in white, rod pigment in black) represent the approximate colors of the pigments. The spectra were obtained by microspectrophotometry of small retinal areas (see Brown and Wald, 1963). Figure prepared by P. K. Brown.

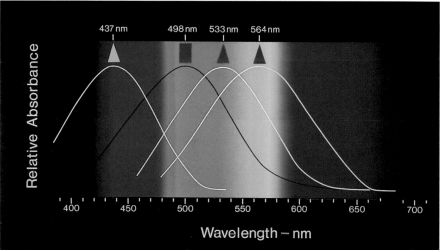

known to generate action potentials in vivo, although some amacrine cells may show only graded potentials and many amacrine cells generate only one or a few action potentials in response to any light stimulus.

Action potentials may be absent from the responses of cells of the distal retina because these cells have relatively short processes and do not transmit information over long distances. Bipolar cells, for example, are usually no more than 100–150 μm in length from the tips of their dendrites to their terminal endings. Thus, electrotonic spread of the graded potentials is sufficient for information to reach from one end of the cell to the other. Another reason many of the retinal cells may function with graded potentials is that such potentials are capable of discriminating a wider range of signals than can all-or-none events. Barlow noted in his 1953 paper that the retina is capable of making such fine intensity discriminations that it seemed unlikely that this could be accomplished by the transmission of all-or-none impulses by retinal neurons. His insight, before anything was known about the electrical responses of the intrinsic retinal neurons, predicted in a sense the finding of graded potential neurons in the retina.

Another early finding of unusual interest was that many of the distal retinal neurons respond to retinal illumination only with hyperpolarizing potentials (Figure 4.1). In action potential-generating neurons, hyperpolarizing potentials are usually associated with inhibition; that is, they drive the membrane potential away from the action potential-generating threshold level (Eccles, 1964). In the retina, however, action potentials are not fired by the distal cells, and presumably excitation can be signaled by hyperpolarizing potential changes. All vertebrate photoreceptor cells only hyperpolarize when illuminated; but many other neurons along the visual pathway are clearly excited during illumination (that is, they depolarize and fire numerous action potentials). These findings constitute compelling evidence that excitation can be signaled by hyperpolarizing potentials in the distal retina.

The intracellular responses shown in Figure 4.1 are typical recordings from a number of different cell types in the mudpuppy retina (Werblin and Dowling, 1969). These responses were assigned to their respective cell types on the basis of intracellular staining, and they were elicited from the cells using a spot of light (100-μm diameter), a small annulus (500-μm diameter), or a large annulus (1-mm diameter). Each cell responds differently to each of these three types of stimuli. The receptors respond best to a spot of

4.1 Intracellular recordings from the mudpuppy retina showing the difference in response of a given cell type to a 100-μm spot and to annuli of 0.5 and 1.0 mm in diameter, all at the same intensity. Note that the distal retinal cells respond to light with sustained, graded, and mainly hyperpolarizing responses.

The receptor, probably a rod, has a relatively narrow receptive field, so the annular stimuli evoke very little response.

The horizontal cell, on the other hand, responds over a much larger area, so relatively large hyperpolarizing responses are recorded with all three stimuli.

The hyperpolarizing bipolar cell responds by hyperpolarizing when the center of its receptive field is illuminated (*left column*). With central illumination maintained (*right column*), annular illumination antagonizes the sustained polarization elicited by central illumination, and a response of opposite polarity is observed in the cell. The annulus used to evoke the response shown in the middle column stimulates both the center and surround region of the bipolar cell receptive field, and a mixed response is recorded.

	SPOT	SMALL ANNULUS	LARGE ANNULUS

Rod photoreceptor cell

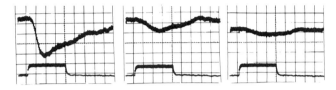

Horizontal cell

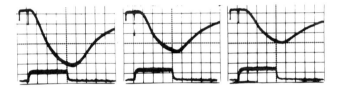

Hyperpolarizing bipolar cell

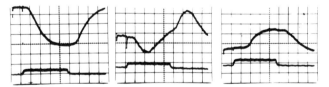

light and only weakly to annuli. The horizontal cells respond with large hyperpolarizing potentials to all three types of stimuli; and the type of bipolar cells illustrated responds with a hyperpolarizing response to a spot of light, a depolarizing response to a large annulus, and a mixed hyperpolarizing–depolarizing response to a small annulus. Thus, with spot and annular illumination it is possible to characterize the responses of each cell type and to describe its respective field organization. Clues as to how the receptive fields are formed can be obtained by considering both the synaptic organization of the cells and the responses to the various stimuli.

Receptors: Light Responses and Interactions
As already noted, all photoreceptors—both rods and cones—hyperpolarize in response to illumination. Tsuneo Tomita and his

4.1 (*cont.*)

The transient amacrine cell gives depolarizing responses at both the onset and cessation of illumination. Its receptive field is somewhat concentrically organized, giving a larger on-response to spot illumination and a larger off-response to annular illumination of 1-mm diameter. With an annulus of 500-μm diameter, the cell responds with large responses at both onset and offset.

The on–off ganglion cell gives bursts of impulses at both the onset and offset of illumination. Its receptive field is similar to that of the transient amacrine cell.

The on-center ganglion cell responds with a sustained depolarization and a maintained discharge of action potentials to spot illumination. With central illumination maintained, large annular illumination (*right column*) inhibits impulse firing for the duration of the stimulus. The small annulus (*middle column*) elicits a brief depolarization and discharge of spikes at onset and a brief hyperpolarization and inhibition of spikes at offset.

The action potentials in these records were reduced in size by the recording conditions. Thus, the underlying slow potentials are larger relative to the size of the action potential than is normally the case.

Modified from Werblin and Dowling (1969), with permission of the American Physiological Society.

Transient amacrine cell

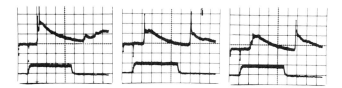

On–off ganglion cell

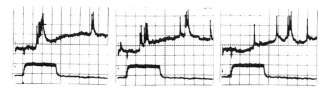

On-center ganglion cell

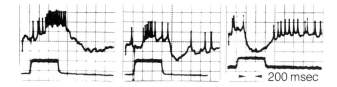

200 msec

colleagues were the first to show unequivocally that vertebrate photoreceptors respond in this way, and Tomita's 1965 paper is a landmark in retinal physiology. Figure 4.3 shows intracellular responses of mudpuppy rods and cones to 200-msec flashes of different intensities (Fain and Dowling, 1973). The mudpuppy retina has a single type of rod and a single type of cone, which absorb maximally at 525 and 572 nm, respectively (Figure 4.2). The basic waveforms (shapes) of the rod and cone responses in the mudpuppy are the same (Figure 4.3a), with an initial transient potential followed by a plateau and a gradual return to baseline. Typically, rod responses in both the mudpuppy and other species return to the baseline much more slowly than do the cone responses. But rods are significantly more sensitive to light than are cones. A light of 8.6 log quanta/cm²-flash elicits a large response in the mudpuppy rod but just a threshold response in the cone (Figure 4.3). When intensity–response curves for rods and cones are compared, rods appear

4.2 Rod and cone responses from the mud-puppy retina.

a Electron micrograph of the outer segment and part of the inner segment of a typical rod (*left*) and cone (*right*).

b Spectral sensitivity curves, showing that rods in the mudpuppy absorb maximally at 525 nm; cones, at 572 nm.

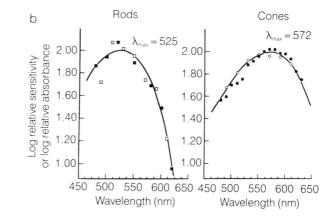

4.3 Both rods and cones respond to light with hyperpolarizing responses (**a**). Rods are more sensitive to light than are cones, however, and their responses are somewhat slower and longer lasting. The responses of the rod were evoked with flashes of 510 nm, and those of the cone with flashes of 610 nm. The number to the left of each response gives the equivalent intensity of the light flash used to evoke the response in log quanta/cm² flash.

Corrected voltage–intensity curves (**b**) show that rods are about twenty-five times more sensitive to light than are cones; that is, the rod curve lies about 1.5 log units further to the left on the intensity axis than does the cone curve. Modified from Fain and Dowling (1973), with permission; copyright 1973 by the AAAS.

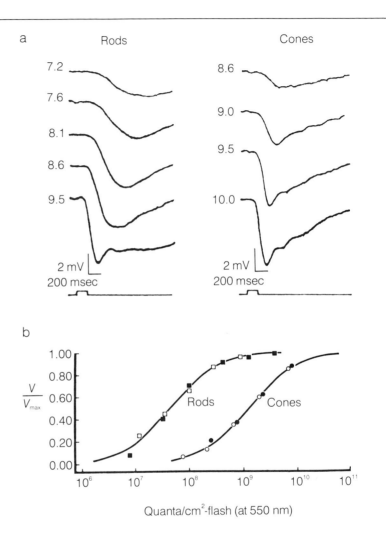

to be about 1.5 log units more sensitive than the cones. Some of this difference can be accounted for by a difference in visual pigment density in the two types of receptors. But when these corrections are taken into account, the sensitivity of the two receptors still differs by over a log unit or about twenty-five times (Figure 4.3b). Thus, rods appear to have a larger intrinsic gain (magnitude of response per quantum absorbed) than do cones.

The intensity–response curves for both rods and cones (Figure 4.3) can be fitted with the Michaelis-Menten equation (see Naka and Rushton, 1966; Baylor et al., 1974), which for photoreceptors is usually written

$$V = \frac{V_{max}I^a}{I^a + \sigma^a}$$

where V is the peak amplitude of the response at intensity I, V_{max} is the peak amplitude at saturation, σ is the intensity necessary to give a response of 0.5 V_{max}, and a is the slope of the function. When $a = 1.0$, this equation predicts that at dim light intensities ($I \ll \sigma$) photoreceptor responses will be linearly related to light intensity and will follow the equation

$$V = \frac{V_{max}}{\sigma}I$$

Near threshold, the gain of a photoreceptor will be the V_{max}/σ. The gains of rods and cones in the linear regions of their responses can be evaluated by comparing the values for V_{max}/σ. But because maximum amplitudes are similar, the gains also can be compared simply by comparing σ. In a number of species, σ has been determined for rods. These data indicate that about 30 quanta absorbed per receptor are needed to give a half-maximal response (Hagins et al., 1970; Dowling and Ripps, 1972). For cones the data for σ are more variable, ranging from about 600 quanta per receptor in the turtle (Baylor and Hodgkin, 1973) to about 3,000 quanta in the primate (Baylor, 1987). In all cases, however, the value of σ is at least a log unit less for rod responses than for cone responses.

Because the maximum amplitudes of rod and cone responses are between 15 and 30 mV and because about 30 quanta are needed to generate a half-maximal response in a rod (σ), near threshold (in the linear region of the receptor response) each quantum should generate a potential change in a rod of 0.5 to 1 mV. Such a potential is large enough to be measured by intracellular recordings; but when such quantal responses were searched for in the toad retina, they were not observed (Fain, 1975b). Rather, when the retina was stimulated with dim flashes such that only about two-thirds of the rods in the illuminated field were absorbing a quantum, responses were recorded every time in the penetrated rod (Figure 4.4). When the light flash was dimmed further, so that even fewer rods caught a quantum, smaller responses were now recorded in the penetrated cell. These experiments showed clearly that receptors must be directly coupled to one another, presumably via the interreceptor gap junctions. Thus, a toad rod is able to respond even when it has not absorbed a quantum (Figure 4.4), because it receives a signal from nearby receptors. It has been estimated that a single rod in the toad

4.4 Intracellular responses of toad rods to flashes of light that bleached fewer than one rhodopsin molecule per receptor. Each column presents a series of consecutive responses to 9-msec flashes of light given at 6-sec intervals. Mean intensities of the flashes (in units of rhodopsin molecules bleached per flash) are given above each column. Arrows indicate the onset of illumination. From Fain (1975b), reprinted with permission; copyright 1975 by the AAAS.

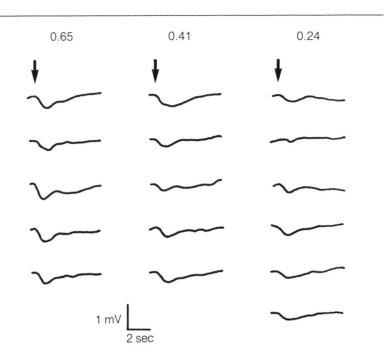

retina may receive signals from as many as 8,000–9,000 rods distributed over an area of 0.5 mm^2 (Fain et al., 1976).*

The first physiological evidence for direct interactions between receptors was in the turtle, where it was shown that spots of light positioned 50 μm away from a cone enhanced its light response (Baylor et al., 1971). Rods in the turtle also receive signals from other receptors, but over a larger area—up to 500 μm in diameter. The direct electrical coupling between photoreceptors appears primarily to link responses of cells of the same spectral class (that is, red rods to red rods, green cones to green cones, and so on), but there is some evidence for weak coupling between different receptor types as well (Schwartz, 1975). As noted in Chapter 3, the extent of gap junctional area observed between receptors in different species and between different types of receptors in the same species varies and presumably reflects the degree of receptor coupling. Also

*Quantal responses have since been recorded from the outer segments of rods with suction pipettes (Baylor et al., 1979b). An outer segment is drawn into a suction pipette and light-evoked currents measured. Such recordings select for currents generated across the outer segment membrane. Furthermore, the membrane potential of the photoreceptor, which is strongly influenced by receptor coupling (Figure 4.4), has little effect on photocurrent generation (Baylor and Nunn, 1985); so, because photocurrent responses recorded from the outer segment are relatively unaffected by the coupling between rods, quantal events can be detected.

as noted in Chapter 3, there is some evidence for chemical synapses between receptors (Lasansky, 1973), but the significance of these junctions is unknown (see Normann et al., 1984).

The role of the extensive electrical coupling between photoreceptors of the same spectral type is not well understood (see discussions in Gold, 1979; Detwiler et al., 1980; Attwell et al., 1984) but is thought to relate primarily to the physiology of the photoreceptors rather than as a pathway for information flow or processing in the retina. For example, isolated photoreceptors are intrinsically noisy electrically; that is, the membrane potential fluctuates significantly in the dark. Electrical coupling, by averaging responses from many rods, can reduce the membrane noise considerably (Lamb and Simon, 1976), improve the signal to noise ratio, and perhaps reduce the number of synapses required for the detection of light responses from receptors (see Falk and Fatt, 1972; Gold, 1979). Electrical coupling between photoreceptors may also serve to increase the amplification of signals across the photoreceptor synapse. In the dark-adapted turtle retina the largest signal amplification between cones and horizontal cells occurs with dim lights, which produce small responses in the receptors (Normann and Perlman, 1979b). Thus, horizontal cell responses will be larger when many receptors with small responses are providing input to a horizontal cell than when a few receptors with larger amplitude responses are providing input to the cell. Electrical coupling, by averaging responses between receptors, may maximize photoreceptor synaptic gain.

Whatever its role, it might appear at first glance that the extensive coupling between receptors, especially rods, would compromise visual acuity significantly. But a comparison of calculated responses of rods surrounding a tiny spot of light with the distribution of light scatter from that spot shows the decline of the two functions over distance to be surprisingly similar (Schwartz, 1976). Thus, receptor coupling, even between rods, may have only a small smudging effect on the resolution by photoreceptors of images falling on the retina. This conclusion is supported by the recordings illustrated in Figure 4.1; the receptor, which was probably a rod, had a considerably smaller receptive field than the other cells shown. It responded well to the spot, but only minimally to the small annulus, and hardly at all to the large annulus. Indeed, the relatively small size of the receptive fields of the photoreceptors has been observed in many species and is often used as a criterion for a photoreceptor recording (Tomita, 1965, 1970).

There is a second synaptic interaction involving the receptors that almost certainly has significance for information processing within the retina. This interaction, again first shown in turtle cones, is an inhibitory feedback from the horizontal cells (Baylor et al., 1971). Such feedback has been demonstrated between horizontal cells and cones in a number of species, although in some cones it is very difficult to detect (Fain, 1975a). Feedback from horizontal cells onto rods has not been found, the only exception being the rods of gecko, which, however, have a number of conelike properties and are believed to be "transmuted" cones (Walls, 1942). The effect of the feedback on the cone photoreceptor response to a step of light is seen as a rapid partial repolarization of the membrane following the initial response. Thus, the cone response typically shows an initial transient component; but this is absent when the feedback is minimal. Because horizontal cells have a much larger receptive field than do the receptors (Figure 4.1), full-field stimulation maximizes horizontal cell responses and feedback, and spot illumination minimizes it. Therefore, the initial transient component in the receptor response is much more obvious in responses elicited with full-field stimulation than in responses elicited with spots of light (Baylor et al., 1971).

Figure 4.5 shows recordings from gecko photoreceptors before and after a pharmacological treatment of the retina (application of aspartate) that eliminated feedback from the horizontal cells by depolarizing and thereby inactivating them (Kleinschmidt and Dowling, 1975). Following aspartate treatment the receptors had an identical light sensitivity and they adapted perfectly normally (Chapter 7). But the waveform of their light response was distinctly different—after aspartate treatment their response to light showed no initial transient component.

Rod receptors that appear not to receive horizontal cell feedback also show an inital transient component in their light responses. These initial transients may arise from voltage-sensitive conductances that are present in the inner segment of the receptor cell and can alter the shape of the photoreceptor response (Fain et al., 1978; Baylor et al., 1979a; Attwell and Wilson, 1980).

How is the basic hyperpolarizing response of the receptors generated? In the dark, the membrane of the outer segment is leaky to Na^+, causing the cell to be partially depolarized (Hagins et al., 1970); typical membrane potentials recorded from photoreceptors in the dark are about -30 mV. Light decreases the conductance of the outer segment membrane, thereby hyperpolarizing the cell

4.5   Intracellular responses from dark-adapted gecko photoreceptors elicited with full-field flashes. The responses shown on the left are those of a control preparation; the responses shown on the right are from a preparation treated with aspartate to block horizontal cell feedback onto the receptors. The sensitivity of the photoreceptor was unaltered by the aspartate treatment, as were the responses to dim flashes. But the waveform of the responses to bright flashes were significantly altered; the responses were simpler in shape and showed no trace of an initial transient. Flash intensity is indicated by the number to the left of each trace. Log I refers to the attenuation by neutral density filters of the full intensity; log I = 0 when no neutral filter used, full intensity corresponds to about 1.3 milliwatts per square centimeter (mw/cm^2). Modified from Kleinschmidt and Dowling (1975), with permission of the Rockefeller University Press.

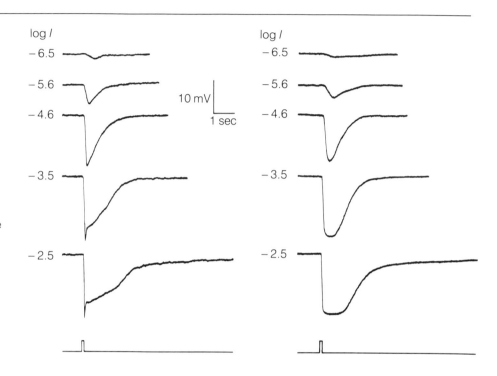

(Tomita, 1970). In response to bright light, the cells may hyperpolarize by as much as 20 to 30 mV; thus, following a bright flash the peak potential of the photoreceptor may be at about − 60 mV, close to the resting potential of most neurons. The final waveform and amplitude of the photoreceptor response depends on feedback from the horizontal cells (cones), or on the activation or inactivation of voltage-sensitive channels (rods) in the inner segment membrane, or on both (Baylor et al., 1971, 1979a; Fain et al., 1978; Attwell and Wilson, 1980; Bader et al., 1982). (See Chapter 7 for a more detailed description of photoreceptor transduction mechanisms.)

Vertebrate photoreceptors thus behave as if darkness is the stimulus, because they are relatively depolarized in the dark, and light turns them off; that is, Na$^+$ channels are shut down and the cell hyperpolarizes. Why vertebrate photoreceptors operate this way is not clear; it may relate to the process of synaptic transmission by the receptors (see Chapter 5). However, with the realization that the receptors are depolarized and release neurotransmitter continuously in the dark, the responses of the second-order cells—the horizontal cells and the bipolar cells—can be better understood.

Horizontal Cells: Electrical Coupling and Color Responses

In the dark, horizontal cells, like the photoreceptors, have membrane potentials of about − 30 mV. In response to light they generally hyperpolarize to levels of − 60 mV or more (MacNichol and Svaetichin, 1958; Tomita, 1963). But, unlike photoreceptors, horizontal cells have very large receptive fields, and, as shown in Figure 4.1, respond with large hyperpolarizing potentials to both spot and small and large annular stimulation. In almost all species the receptive field size of a horizontal cell far exceeds its dendritic spread. In fishes typical dendritic fields are between 30 and 150 μm (Stell and Lightfoot, 1975), whereas receptive field sizes are often 2–5 mm or more (Naka and Rushton, 1967). The very large receptive field size of the horizontal cells is attributable to the large gap (electrical) junctions that exist between the horizontal cell perikarya or processes (Kaneko, 1971a) (see Figure 3.11 and Plate 3). The resistances of these junctions are very low (∼ 30 mΩ) relative to the resistance of the cell (∼1,000 mΩ). So when current is injected into an individual horizontal cell, virtually all of the current will flow into the neighboring cells (Lasater and Dowling, 1985a,b). This electrical coupling, and consequently the receptive field size of horizontal cells in fishes, can be modulated by synaptic input from the interplexiform cells (see Chapter 5).

Figure 4.6 shows horizontal cell responses recorded intracellularly from the all-rod retina of the skate over a wide range of intensities (Dowling and Ripps, 1971). From threshold light levels the responses grow in amplitude with increasing intensity and reach a maximum amplitude with light flashes about 4 log units above threshold (Figure 4.6b). Thereafter, the responses remain constant in amplitude but become greatly prolonged. Rod-related horizontal cell responses in most species, like the responses of the rods themselves, recover slowly from bright flashes; horizontal cell responses deriving from cones, on the other hand, usually recover rapidly after the cessation of a light stimulus, regardless of intensity (see, for example, Nelson et al., 1976). Horizontal cell responses can also be fitted with the Michaelis-Menten equation. But in the case of many horizontal cells the intensity range over which they respond is somewhat broader than that for the receptors; that is, the slope (a) of the function is 0.7–0.8 for horizontal cells rather than 1.0 (Figure 4.6b). How the horizontal cells manage to respond over a greater intensity range than the receptors do is not clear; the difference presumably relates to the nature of the synaptic interaction between the receptors and the horizontal cells.

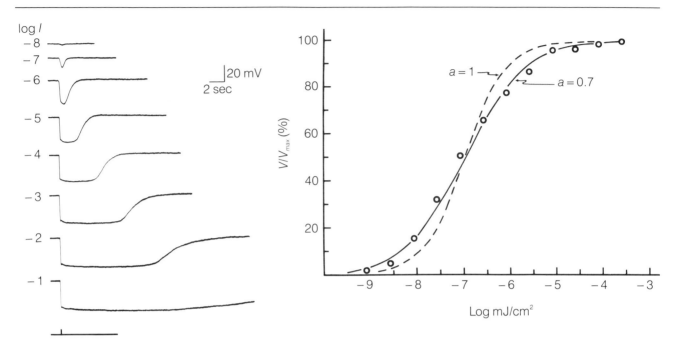

4.6 **a** Horizontal cell responses recorded from the dark-adapted skate retina. Flash duration was 0.2 sec, and the log *I* value is the density of the neutral filter interposed in the test beam (when log *I* = 0, retinal irradiance was 1.22 mw/cm²). With increasing stimulus intensity the responses grew in amplitude and their duration increased.

b Average voltage–intensity curve for skate horizontal cells. The horizontal cells (*solid line*) respond over a broader intensity range than do the receptors (*dotted line*). The value of *a* in the Michaelis-Menten equation, to which the data is fitted, is about 0.7 rather than 1, which is the value of *a* for the receptors. From Dowling and Ripps (1971), reprinted with permission of the Rockefeller University Press.

Most horizontal cells in all species so far examined only hyperpolarize to illumination, regardless of intensity or wavelength, and they are called luminosity or L-type horizontal cells (Figure 4.6). But animals that have highly developed color-vision mechanisms (for example, fishes and turtles) have some horizontal cells that respond to some wavelengths of light with hyperpolarizing potentials and to other wavelengths with depolarizing potentials (Mac-Nichol and Svaetichin, 1958; Tomita, 1965; Fuortes and Simon, 1974; Leeper and Copenhagen, 1982). These cells are called chromaticity or C-type horizontal cells. Two basic types of responses from C-type horizontal cells are commonly seen: biphasic re-

4.7 Three types of cone-driven horizontal cell responses recorded in the carp retina.

a Response of an L-type cell, which hyperpolarizes to light of all wavelengths.

b Response of a biphasic C-type cell, which hyperpolarizes to short wavelengths of light and depolarizes to long-wavelength stimuli.

c Response of a triphasic C-type cell, which hyperpolarizes to short and long wavelengths and depolarizes to intermediate wavelengths. From Toyoda et al. (1982), reprinted with permission of Alan R. Liss.

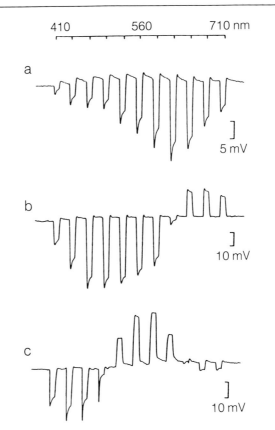

sponses, in which the cells hyperpolarize to green light and depolarize to red light (R/G cells) or hyperpolarize to blue light and depolarize to green light (G/B cells); and a triphasic response, in which a red–green–blue cell hyperpolarizes to red and blue lights but depolarizes to intermediate wavelengths of light (Figure 4.7). In carp and goldfish retinas the monophasic L-type cells and the biphasic and triphasic C-type horizontal cells each have a distinctive morphology (Figure 2.6), and their connections to the receptors have been traced (Stell and Lightfoot, 1975). Each cell type contacts a different subset of cones in a unique way, a finding suggesting that the C-type responses are established by a series of feedback and feedforward connections between the horizontal cells and the spectrally different cones (Stell et al., 1982; Toyoda et al., 1982).

Evidence, both anatomical and physiological, indicates that the hyperpolarizing light responses of all horizontal cells are generated in the same basic way—by a decrease in transmitter release from the receptors (Chapter 5). Although direct evidence is lacking, the presumption is that the depolarizing responses in the C-type cells result from an increase in transmitter release from the receptors, perhaps as a result of increased depolarization of the receptor terminals caused by withdrawal of hyperpolarizing horizontal cell transmitter during light stimulation.

Bipolar Cells: Center–Surround Organization

Bipolar cells, like receptors and horizontal cells, respond to light with sustained graded potentials (Figure 4.1). In all species, however, two types of bipolar cells have been described: those that hyperpolarize in response to illumination of the center of the cell's receptive field and those that depolarize in response to center illumination (Werblin and Dowling, 1969; Kaneko, 1970; Matsumoto and Naka, 1972; Schwartz, 1974; Dacheux and Miller, 1981). The former cells are called hyperpolarizing or off-center bipolar cells (Figure 4.1), and the latter are called depolarizing or on-center bipolar cells. Furthermore, with almost no exceptions the bipolar cell receptive field is concentrically organized into antagonistic zones, such that illumination of the surround decreases the response to illumination at the center of the receptive field. Evidence indicates that the center response reflects direct receptor–bipolar synaptic interaction, whereas the surround response is mediated by horizontal cell activity.

In some species surround or annular illumination does not drive the membrane potential beyond the resting potential (Werblin and Dowling, 1969). Thus, to observe the effects of the antagonistic surround, illumination of the receptive field center must be present (Figures 4.8 and 4.9). In other species annular illumination alone will evoke a response opposite in polarity to that evoked by a center spot (Kaneko, 1970). The size of the antagonistic surround is very much larger than the size of the central field of the bipolar cells. In the mudpuppy, the center of the receptive field is about 100 μm in diameter, and an annulus with a diameter of about 1 mm is most effective in eliciting the antagonistic surround response (Figure 4.1). In other cold-blooded species, the size of bipolar cell center ranges from 100–200 μm in diameter (goldfish; Kaneko, 1973) to about 400 μm (tiger salamander; Werblin, 1977b); these cells also have correspondingly large surrounds (1–2 mm diameter). With a smaller annulus (Figure 4.1) or with a full-field illumination, both the center and surround mechanisms are stimulated. The center

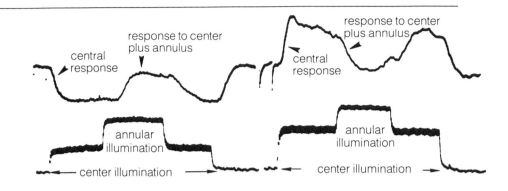

4.8   Records from mudpuppy bipolar cells showing their antagonistic center–surround receptive field organization. *Left,* Center hyperpolarizing cell; *right,* center depolarizing cell. In each experiment, the center illumination was maintained while an annulus of 500-μm diameter was presented. The responses to center illumination in each case was antagonized by the annular illumination. From Werblin and Dowling (1969), reprinted with permission of the American Physiological Society.

response has a shorter latency than the surround response, so a mixed, sawtooth-shaped response is elicited in the cells (Figure 4.1).

The central receptive field of some bipolar cells has been found to be larger than the dendritic field of the cells. The large central receptive field size of these bipolar cells, like that of horizontal cells, appears to be due to electrical coupling between the cells (Kujiraoka and Saito, 1986). The electrical coupling between bipolar cells is not nearly as extensive as that between horizontal cells, however; thus, the absolute size of the receptive field center of bipolar cells is always very much smaller than is the receptive field size of horizontal cells.

The difference in latencies as well as the antagonistic interactions between center and surround mechanisms in bipolar cells are shown in Figure 4.9. In this experiment the retina was stimulated by an annular flash (250-μm radius) that was sufficiently bright to stimulate the center of the field with scattered light. When the center of the field was illuminated by scattered light only (no maintained illumination of the field center, top trace), the response reflected center stimulation only, that is, a pure hyperpolarizing response was elicited. As maintained illumination of the center increased, the surround mechanism became apparent as a turning off of the center response after a delay, and a mixed center–surround response resulted. With the center of the receptive field maximally hyperpolarized by bright, maintained, central illumination, the annular flash evoked a pure surround response; that is, the cell only depolarized.

Dark membrane potentials for hyperpolarizing bipolar cells are typically −30 mV, whereas for depolarizing bipolar cells somewhat larger resting potentials have been reported (up to −50 mV) (Ash-

4.9 Records from mudpuppy bipolar cells showing the latency differences and interactions between center and surround responses. The series of responses was elicited by annular stimulation sufficiently bright to stimulate the center of the field by scattered light. Maintained central illumination, at first absent (*top trace*), was added and gradually increased in intensity to reduce the effects of the light scatter from the flashing annulus into the receptive field center. The effective ratio of center and surround stimulation was shifted as central illumination intensity increased in favor of surround response. Note, however, that the hyperpolarizing center response always occurred earlier, even though the surround response became dominant (*lower traces*). From Werblin and Dowling (1969), reprinted with permission of the American Physiological Society.

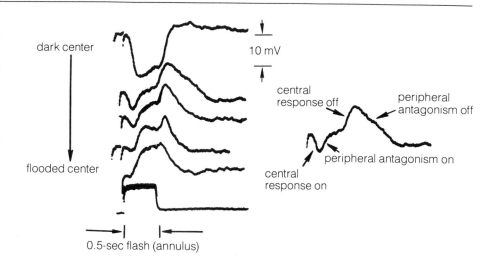

more and Falk, 1980). Both cell types give center responses of 20–25 mV, or occasionally even larger (Schwartz, 1974; Ashmore and Falk, 1980; R. F. Miller, personal communication). The graded center responses of the bipolar cells can be fitted by the Michaelis-Menten equation, but for bipolar cells the slope of the function is much steeper than that of either the receptors or horizontal cells. Whereas the slope of the Michaelis-Menten function is 1.0 for receptors, and 0.7–1.0 for horizontal cells, it is approximately 1.4 for bipolar cells (Werblin, 1974). This means, of course, that the central field of the bipolar cells responds over a narrower range of intensities than do the fields of receptors and horizontal cells (~2 log units in bipolar cells and 4 log units in horizontal cells). But surround illumination, by antagonizing the central response and driving the potential back toward the dark potential, extends the intensity domain of the bipolar response; that is, the bipolar cell responds to large-field illumination over a broader range of intensities than it does to small-field stimulation. This effect can be illustrated by examining the amplitude of the plateau potential that follows the initial center response when both center and surround are illuminated together. This same effect is shown in a different and more dramatic way in Figure 4.10. The graded center response was measured in the presence of three annuli of different intensities. The position of the relatively narrow bipolar cell intensity–response curve to spot illumination shifted with the intensity of the

4.10 Voltage–intensity curves for the center response of a depolarizing bipolar cell in the tiger salamander retina recorded in the presence of annuli of three different intensities. Note that the antagonistic interactions of the center and surround mechanisms shift the operating curves of the bipolar cell to the right on the intensity axis. From Werblin (1977), reprinted with permission of Academic Press.

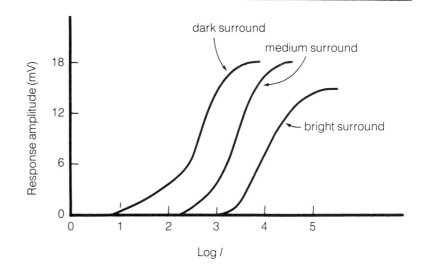

surround illumination such that, with the three surrounds, the bipolar cell responded over the range of intensities of a horizontal cell. Thus, bipolar cell responses, by virtue of their center–surround antagonistic organization, reflect not only contrast detection but also a retinal network adaptive mechanism. That is, the antagonistic surround permits the bipolar cell to adjust its operating range on the basis of the background or surround illumination. (Other retinal light and dark adaptation mechanisms will be discussed in Chapter 7.)

Bipolar cells also may play an important role in color vision mechanisms in some species (Kaneko, 1973; Yazulla, 1976; Kaneko and Tachibana, 1983). In fishes and turtles some bipolar cell responses are color-coded. There are bipolar cells with centers that respond best to red light and surrounds that respond best to green light (single-opponent cells); other bipolar cells have center and surround responses that change polarity when wavelength changes (double-opponent cells). Finally, some bipolar cells appear to have a mix of single- and double-opponent fields. Two types of double-opponent cells have been found in the carp retina. One type has a center response of hyperpolarization to red light and depolarization to blue-green light; its surround response is depolarization to red light and hyperpolarization to blue-green light (Figure 4.11). The other type is its mirror image; red light depolarizes whereas green light hyperpolarizes the center mechanisms, and red light hy-

4.11 Responses of a double-opponent bipolar cell in the carp retina. A spot of red light hyperpolarized the cell, whereas an annulus of red light depolarized it. Conversely, a spot of blue-green light primarily depolarized the cell, whereas an annulus of the same wavelength hyperpolarized it. Modified from Kaneko and Tachibana (1983), with permission of Pergamon Journals Ltd.

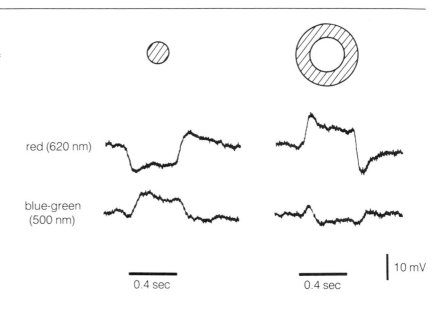

red (620 nm)

blue-green
(500 nm)

0.4 sec   0.4 sec   10 mV

perpolarizes whereas green light depolarizes the surround mechanism. It has been proposed that single-opponent cells in visual systems may play a role in successive color-contrast phenomena (Daw, 1973); that is, following the sustained viewing of a red field, subsequent viewing of a blank piece of white paper results in the appearance of a greenish field and vice versa. The double-opponent cells, on the other hand, may underlie simultaneous color-contrast phenomena; that is, when a gray field is surrounded by a red background, the gray field looks greenish, and vice versa.

Finally, some bipolar cells that receive their input exclusively or primarily from rods have been described in the carp and the dogfish (Saito et al., 1979; Ashmore and Falk, 1980). These cells are primarily depolarized by centered spots of light; and they are significantly more sensitive (by over a log unit) to photic stimulation than are the cone-dominated bipolar cell responses. The waveform of the rod bipolars is likewise different, consisting of an initial transient component followed by a plateau that continues after the termination of a bright stimulus. Cone bipolar cells tend to have a much squarer response to centered illumination (Figure 4.1), and their response terminates promptly upon cessation of the light stimulus. In the carp, surround responses of the rod bipolar cells were readily demonstrated, but in the dogfish surround responses were difficult to see clearly, although hints of these responses were present. In the cat, bipolar cells identified as rod bipolar cells were always hyperpolarizing cells, their response waveform was quite square, and no antagonistic surround was detected (Nelson and Kolb, 1983). These observations are at variance with those from the cold-blooded vertebrates, and their significance is unclear. In the rabbit rod bipolar cells depolarize to central field illumination and their responses closely resemble those of rod bipolar cells in the cold-blooded vertebrates (Dacheux and Raviola, 1986). Furthermore, rod bipolar cells in the rabbit are believed to have an antagonistic surround response (Dacheux and Raviola, 1986).

Amacrine Cells: Transient or Sustained Responses

In most retinas two basic types of amacrine cell responses—transient and sustained—have been observed, although in most retinas many more transient amacrine cell responses than sustained amacrine cell responses are recorded (Kaneko, 1971b; Toyoda et al., 1973; Naka and Ohtsuka, 1975). In some retinas, such as the mudpuppy retina, there may be very few, if any, amacrine cells with sustained responses (Dowling and Werblin, 1969).

More is known about the transient amacrine cells than about the sustained ones. Transient amacrine cells usually give on- and off-responses to illumination presented anywhere in their receptive fields (Figure 4.12a). They always respond by depolarizing and are the first neurons along the visual pathway to respond primarily with transient and depolarizing potentials. Some differences between the on- and off- components of the transient amacrine responses may be observed, however; and these differences can occasionally be related to the geometry or position of the stimulus used. For example, in Figure 4.1, the transient amacrine cell responded with a large on-response and a small off-response to illumination by a central spot. With small annular illumination the on- and off-responses were comparable; whereas with large annular illumination the off-response was more prominent. Some transient amacrine cells in some species respond only with on-responses regardless of stimulus configuration, whereas others respond only with an off-response. Transient amacrine cells generally do not show a center–surround antagonistic receptive field organization.

Action potentials are often seen superimposed on the depolarizing on- and off-responses of the transient amacrine cell (Dowling and Werblin, 1969). In many cells only one or two spikes are observed riding on the transient depolarizations regardless of intensity or configuration of the stimulus. Figure 4.13 shows the effect of

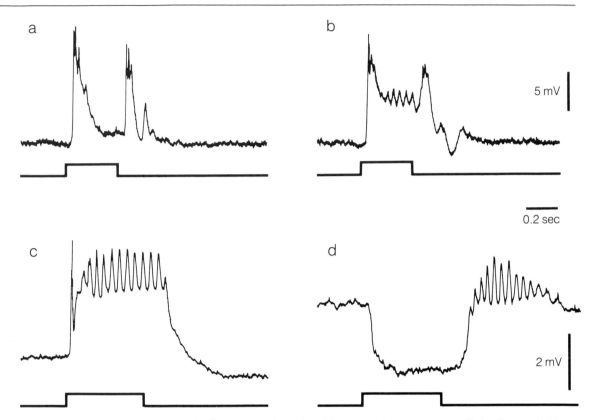

4.12 Responses of transient (**a**) and sustained amacrine cells (**c, d**) recorded in the catfish retina. Some amacrine cells (**b**) show a mix of transient and sustained components. Regenerative potentials are often observed superimposed on the transient depolarizations, whereas oscillatory potentials are frequently observed superimposed on the sustained potentials. Unpublished records from M. Sakuranaga and K.-I. Naka.

intensity on the on-component of a transient amacrine cell response. With increasing intensity, the response grew rapidly in amplitude and the latency of the response shortened considerably. But even at saturation only two action potentials were generated. Thus, it is likely that the slow potential response is the more important component for signal transmission by the amacrine cells. Further, amacrine cell action potentials may be of two types; large spikes generated within the soma, and small spikes generated within the processes of the cell (Miller and Dacheux, 1976a). This finding raises the question of whether amacrine cell spikes are actively

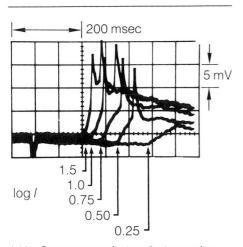

4.13 On-response of a transient amacrine cell to light flashes of increasing intensity in the mudpuppy retina. With increasing intensity, latency of the response decreased and response amplitude increased. Even at the brightest intensity (log I = 1.5), only two regenerative potentials were observed superimposed on the graded potential. From Werblin and Dowling (1969), reprinted with permission of the American Physiological Society.

propagated along amacrine cell processes or only serve as a local amplifying mechanism for amacrine cell potentials (Miller, 1979).

Transient amacrine cells have been reported to have resting potentials ranging from -30 to -60 mV, with somatic spikes up to 100 mV riding on 20- to 30-mV transient depolarizations (Werblin and Dowling, 1969; Toyoda et al., 1973; Miller, 1979). As the data in Figure 4.13 indicate, transient amacrine cells also have an extremely narrow intensity–response function. Indeed, 90 percent of the response range of a transient amacrine cell is generated by about 1 log of intensity (see also Figure 6.18). Transient amacrine cells are also very responsive to moving stimuli. When a spinning windmill stimulus is presented to a transient amacrine cell, the cell rapidly depolarizes and remains depolarized until the windmill stops, at which point the potentials promptly return to rest (Werblin, 1972; Werblin and Copenhagen, 1974).

Figure 4.12 shows examples of sustained amacrine cell responses (Sakuranaga and Naka, 1985b,c). Both depolarizing (Figure 4.12c) and hyperpolarizing (Figure 4.12d) sustained amacrine cell responses have been described. In both responses spikelike and oscillatory potentials are often superimposed on the depolarizing components (Hosokawa and Naka, 1985). The origin of the oscillations is not known; they may possibly represent partial regenerative events. The responses of sustained amacrine cells also may be color-coded; that is, the polarity of the response may depend on wavelength (Kaneko, 1971b; Toyoda et al., 1973). In addition, a number of the sustained amacrine cells show center–surround antagonism (Toyoda et al., 1973). Finally, amacrine cell responses that have both transient and sustained components have been described (Matsumoto and Naka, 1972; Norton et al., 1970; Toyoda et al., 1973). Figure 4.12b shows such a response: a sustained depolarization with superimposed oscillations occurs between the transient on- and off-responses.

In a number of respects, the responses of the sustained amacrine cells resemble those of the distal retinal neurons, especially those of the horizontal and bipolar cells (Sakuranaga and Naka, 1985b). The responses of transient amacrine cells, on the other hand, are quite different. Figure 4.14 shows an example of this. In these experiments the retina was stimulated with a sinusoidal light stimulus of increasing frequency. The responses of horizontal and sustained amacrine cells follow reasonably well the sinusoidal input, with response amplitudes decaying monotonically with increasing fre-

4.14 Responses of a horizontal cell, a hyperpolarizing sustained amacrine cell, and a transient amacrine cell to a sinusoidal stimulus of increasing frequency in the catfish retina. Both the horizontal cell and sustained amacrine cell followed the sinusoidal stimulus with response amplitudes decaying monotonically with increasing frequency. The transient amacrine cell, on the other hand, responded by sharply depolarizing both to the increases and decreases of illumination (that is, their responses showed frequency doubling and rectification). Furthermore, the transient amacrine cell stopped responding at a much lower frequency than did the horizontal or sustained amacrine cell. The arrow points to sharp depolarizing potentials the sustained amacrine cell generated at certain frequencies. From Sakuranaga and Naka (1985b,c), reprinted with permission of the American Physiological Society.

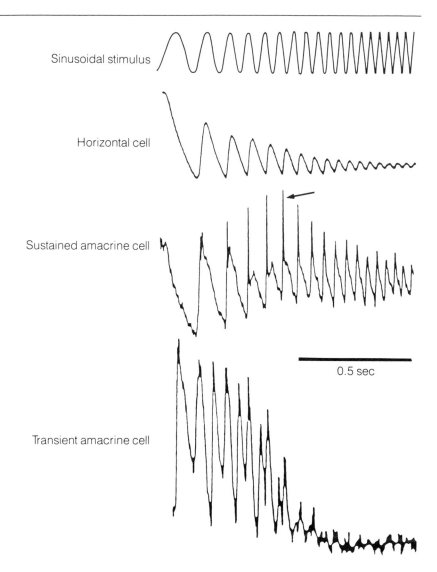

quency. The responses of transient amacrine cells, on the other hand, do not accurately follow such a stimulus. The transient amacrine cells produce sharp depolarizing potentials to both the increase and the decrease of intensity during a single stimulus cycle; that is, they show rectification properties and frequency doubling. Also, the decrease in response amplitude of the transient amacrine cells is not monotonic with increasing frequency, and the cells stop responding at a much lower frequency than the horizontal or sustained amacrine cells do. Note, however, that the sustained amacrine cells also produce sharp depolarizing transient potentials to certain frequencies of the sinusoidal stimulus (arrow in Figure 4.14). This finding indicates that the sustained amacrine cell has some properties of the transient amacrine cells, and in this respect behaves like a transition cell (Sakuranaga and Naka, 1985b).

Interplexiform Cells

Only two reports of intracellular recordings from interplexiform cells have appeared. These recordings were from the retina of the dace, a cyprinid fish; and most of the responses were sustained in nature, like those of sustained amacrines (Hashimoto et al., 1980). More recent recordings suggest that there are both transient and sustained components in some interplexiform cell responses (Y. Hashimoto, personal communication).

Ganglion Cells

There are considerably fewer reports of intracellular ganglion cell recordings than for any other retinal cell type except the interplexiform cell, especially in mammals (but see Saito, 1983). Most reports, from nonmammalian species, describe two basic kinds of ganglion cells: those that give sustained on-center or off-center responses and show an antagonistic surround; and those that respond to flashes of light with transient on–off responses. Both types of ganglion cells are found in the mudpuppy retina and their responses are shown in Figures 4.1 and 4.17. The response of the sustained type of ganglion cell shown in Figure 4.1 is from an on-center cell. With central spot illumination, a sustained slow potential and a steady discharge of spikes was evoked from the cell. With some central illumination maintained, large annular illumination hyperpolarized the cell and inhibited the firing in a sustained fashion.

The other ganglion cell type in the mudpuppy retina gives transient responses at the onset and cessation of stimulation, that is, on–off responses. Differing amounts of on- and off-contributions were evoked with the different stimulus configurations. This kind of response is not seen in all cells of this type; and often similar on- and off-responses are observed regardless of stimulus position, intensity, or configuration. Numerous action potentials are usually associated with the transient depolarizations seen in these ganglion cell responses; and the number of spikes fired appears to be closely related to the amount and extent of depolarization, unlike the sit-

4.15 Intensity–response function for sustained on-center (*top*) and transient on–off (*bottom*) ganglion cells in the mudpuppy retina. Responses were the number of action potentials in a 5-sec period following the test flash. The intensity–response curve for the sustained ganglion cell is much broader than the intensity-response curve for the transient cell. From Thibos and Werblin (1978), reprinted with permission of the Physiological Society.

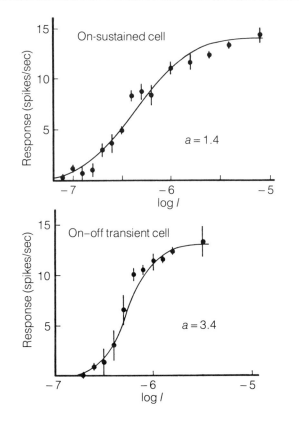

uation with the transient amacrine cells discussed earlier. A number of on–off cells in the mudpuppy are highly sensitive to movement and some show direction-sensitive responses (Werblin, 1970).

As can be seen from Figure 4.1, the on–off ganglion cell resembles the transient amacrine cell in terms of the transient nature of its responses and receptive field organization, whereas the sustained ganglion cell resembles the bipolar cell in terms of its sustained responses and center–surround antagonistic receptive field organization. Thus, the responses of the two basic types of ganglion cells in the mudpuppy appear to relate to the activity of one or another of the cell classes (that is, amacrine or bipolar cells) providing input to ganglion cells. Further evidence for this is provided by comparing the intensity–response curve for the center responses of sustained and transient ganglion cells in the mudpuppy (Thibos and Werblin, 1978). In both cases, the curves were fit to the Michaelis-Menten equation and values of *a* determined (Figure 4.15). For sustained ganglion cells (three on-center, four off-center)

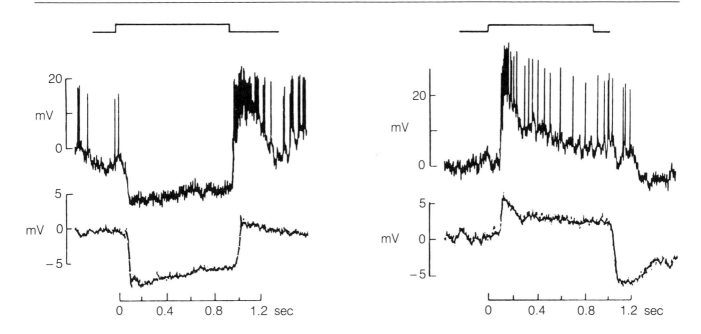

4.16 Intracellular recordings from an off-center ganglion cell (*top records*) and a hyperpolarizing bipolar cell (*bottom records*) in the turtle retina to spot (*left*) or to annular illumination (*right*) of equal intensities (log *I* = −3.6). Modified from Marchiafava and Weiler (1980), with permission of the Royal Society.

a averaged 1.4, in excellent agreement with the value of *a* found for bipolar cells (1.2–1.4). The intensity response curves of transient ganglion cells were much steeper, with *a* equal to 3.4. This means that the responses of the transient ganglion cells rise from 5 to 95 percent of maximum when the intensity increases by 0.8 log units, much as the responses of the transient amacrine cells do.

In the turtle retina also, intracellular recordings show two basic physiological types of ganglion cells (Marchiafava and Weiler, 1980; Marchiafava, 1983). One type gives sustained on- or off-center responses to central illumination, and peripheral illumination antagonizes the center response. The waveform and receptive field organization of these cells closely resemble those of the bipolar cells in turtles (Figure 4.16). The other ganglion cell type gives transient responses to the onset and cessation of illumination, and its response resembles transient amacrine cell responses. Many of these cells are direction-sensitive.

Functional Organization of the Retina

Although there are relatively few reports characterizing the response properties of ganglion cells by intracellular recordings, there are abundant studies on the receptive field properties of ganglion cells made by extracellular recordings (Chapter 2). Possible mechanisms underlying the formation of receptive fields of ganglion cells and other retinal neurons can be derived from data on the intracellular response properties of the retinal cells along with information on the synaptic organization of the retina.

Figure 4.17 is a highly simplified scheme that suggests how some of the potentials and certain of the receptive fields of the retinal neurons may be produced by the synaptic interactions occurring within the retina (Dowling, 1970). The drawing correlates a basic wiring diagram of the retina, based on the connections shown in Figure 3.17, with typical potentials recorded from the mudpuppy retina. The experimental conditions that would give rise to the responses shown in Figure 4.17 are presentation of a bright flash of light to the receptor on the left while both receptors are continuously illuminated with a dim background light. (The background light is needed to demonstrate the effect of the antagonistic synaptic interactions in the mudpuppy.) The cells on the left side of the figure have responded as if stimulated with centered spot illumination, whereas the cells on the right have responded as if stimulated with annular (surround) illumination. Although the responses illustrated in Figure 4.17 are all from the mudpuppy, the following discussion is intended to be more general, and it draws on results from many species.

Outer Plexiform Layer: Bipolar Cell Receptive Fields

Vertebrate photoreceptors respond relatively autonomously (Figure 4.17). Thus, a large response is observed in the receptor stimulated with the bright flash (left receptor) whereas the adjacent receptor that is not illuminated (right receptor) shows only a small response that probably reflects mainly the electrical coupling between the photoreceptor cells. The anatomy of the retina indicates that bipolar and horizontal cells are both activated by the receptors (Chapter 3). The scheme of Figure 4.17 further suggests that bipolar cells are polarized strongly in a graded and sustained fashion by direct contacts between receptor and bipolar cells (left side) and that these bipolar cell potentials are antagonized by horizontal–bipolar cell interactions (right side). Evidence from anatomical

4.17 Summary diagram correlating the synaptic organization of the vertebrate retina with some of the intracellularly recorded responses from the mudpuppy retina. This figure attempts to show how the receptive field organization of the hyperpolarizing bipolar cells, off-center ganglion cells, and on–off ganglion cells is established. The responses occurring in the various neurons upon illumination (*bar*) of the left receptor are indicated.

The hyperpolarizing bipolar cells and off-center ganglion cells (G_1) respond to direct central illumination (*left side*) by hyperpolarizing; to indirect (surround) illumination (*right side*) by depolarizing. Note that the switch from hyperpolarizing to depolarizing potentials along the surround illumination pathway occurs at the horizontal–bipolar junction.

The on–off ganglion cell (G_3) receives strong inhibitory input from amacrine cells; the figure suggests that these cells receive their excitatory input from both amacrine and bipolar cells. Inhibitory feedback synapses from amacrine cells onto the bipolar terminals are also indicated.

R, receptors; H, horizontal cell; B, bipolar cells; A, amacrine cells; G, ganglion cells; + with open circles represents excitatory synapses; − with filled circles represents inhibitory synapses. Modified from Dowling (1970), with permisson of J. B. Lippincott Company.

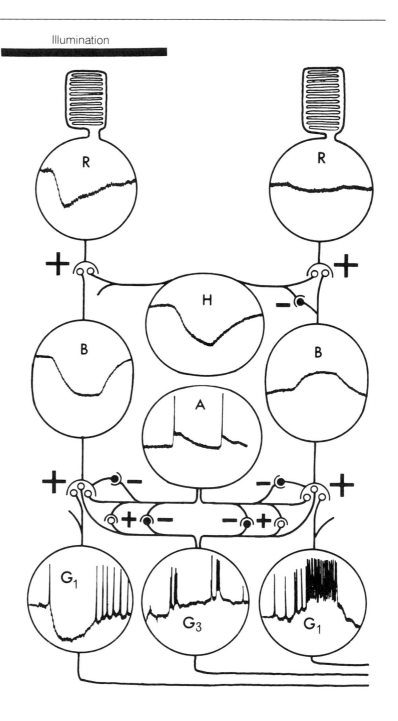

studies suggests that such interactions between horizontal and bipolar cells could occur at direct synapses between horizontal and bipolar cells (shown in Figure 4.17), whereas physiological evidence indicates that horizontal cells may also exert their effects by feeding back onto the receptors (not shown in Figure 4.17) and producing antagonistic bipolar cell responses in that way.

Because horizontal cells have a greater lateral extent in the outer plexiform layer than do the bipolar cells, a center–surround receptive field organization is observed in the bipolar cell response. The center response pathway (left side) appears to be mediated by the direct synapses between receptor and bipolar cells, whereas the antagonistic surround response (right side) reflects a receptor–horizontal–bipolar cell pathway. The surround response in the bipolar cell has a longer latency than does the center response, reflecting the fact that horizontal cell responses are relatively slow and also that additional synapses are involved. Furthermore, in the mudpuppy the physiologically determined receptive field center of the bipolar cell matches closely in diameter the dendritic spread of the bipolar cells, whereas the surround response is very much larger and approximates the receptive field size of the horizontal cells (see Figure 4.1). The only other cell type in the retina spreading far enough to account for the antagonistic surround in the bipolar cell response is the amacrine cell. In the mudpuppy, however, virtually all amacrine cells respond transiently to retinal illumination at both onset and offset. The surround inhibition observed in the bipolar cell response is, on the other hand, graded and sustained and has the approximate form of the horizontal cell response.

Impressive evidence in favor of the hypothesis that horizontal cells form the surrounds of bipolar cells has been provided by Naka and his colleagues, who, following the pioneering work of Maksimova (1970), injected currents into catfish and dogfish horizontal cells (Naka, 1972; Naka and Witkovsky, 1972; Sakuranaga and Naka, 1985a). They found that hyperpolarizing currents injected into the horizontal cells mimic the effects of surround illumination on both bipolar and ganglion cells. In carp and turtle retinas, such experiments also have been carried out with similar results (Toyoda and Tonasaki, 1978; Marchiafava, 1978; Toyoda and Kujiraoka, 1982).

The bipolar cells illustrated in Figure 4.17 are center-hyperpolarizing (left side), surround-depolarizing (right side) cells. Responses elicited in the receptor and in this kind of bipolar cells with centered spots are of the same polarity; therefore, the receptor synapse is indicated in this situation as an excitatory junction (+). The horizontal–bipolar synapse, on the other hand, is indicated as an inhibitory synapse (−), because the responses in the horizontal and bipolar cells are of different polarity. It must be remembered, however, that another class of bipolar cells exists in all retinas— the center-depolarizing, surround-hyperpolarizing cells. The responses of these cells can be accounted for in the scheme of Figure 4.17 by simply reversing the signs of the receptor to bipolar (+ to −) and horizontal to bipolar (− to +) synapses. There is evidence for this arrangement in that hyperpolarizing currents passed into horizontal cells hyperpolarize on-center bipolar cells, whereas the same currents depolarize the off-center bipolar cells (Toyoda and Kujiraoka, 1982). The mechanisms underlying the receptor–bipolar cell synapses will be discussed further in the next chapter.

Inner Plexiform Layer: Ganglion Cell Receptive Fields
The amacrine cell is the first cell along the visual pathway that is exclusive to the inner plexiform layer. Many amacrine cells respond to light mainly with transient depolarizing potentials at the onset and cessation of spot illumination placed anywhere in the cell's receptive field (Figure 4.17). How the sustained responses of the distal retinal cells are converted to transient responses at the level of amacrine cells is not known, but the anatomy of the bipolar– amacrine cell synaptic complex provides the basis for a suggestion. Synapses of amacrine cells are made back onto bipolar terminals just adjacent to the bipolar ribbon synapses and could conceivably turn off bipolar cell excitation locally; a transient response in the amacrine cell could result (Burkhardt, 1972). But feedback alone is unlikely to account fully for the transient responses of amacrine cells, and evidence that the intrinsic membrane properties of these cells shape their responses significantly has been provided (Barnes and Werblin, 1986).

The responses of the two basic types of ganglion cells found in the mudpuppy and other retinas appear to be closely related to the responses of the input neurons to the ganglion cells. The G_1 ganglion cells have a receptive field organization very similar to that of the bipolar cells. Central illumination hyperpolarizes both the bipolar and ganglion cells in a sustained fashion (left side of Figure 4.17), and surround illumination depolarizes the bipolar and ganglion cells in a sustained fashion (right side of Figure 4.17). This type of ganglion cells would appear to receive most of its synaptic input directly from the bipolar cell terminals through excitatory synapses, and there is anatomical and physiological evidence that this is the case (see Chapter 3; Naka, 1976, 1977; Miller and Dacheux, 1976a,b,c). The ganglion cells illustrated in Figure 4.17 are off-center cells; but there are, of course, on-center cells present in the mudpuppy retina and other retinas that have responses and receptive field properties very similar to the on-center bipolar cells. The on-center ganglion cells would appear to receive the bulk of their input from the on-center bipolar cells.

Evidence that the receptive fields of the sustained, on- and off-center ganglion cells (see Figure 2.13) reflect primarily the receptive fields of bipolar cells has been provided by two types of experiments. As already noted, Thibos and Werblin (1978) showed in the mudpuppy that the center responses of the on- and off-center cells have an intensity–response function very similar to that of bipolar cells; furthermore, they showed that the surround response has an intensity–response function very similar to that of the horizontal cells. Second, Caldwell and Daw (1978b) have shown in the rabbit retina that antagonists to inner plexiform layer neurotransmitters, for example, picrotoxin and strychnine, do not change fundamentally the concentric receptive field organization of the sustained on- or off-center cells. It should be emphasized that these experiments do not indicate that the sustained on- and off-center ganglion cells receive input exclusively from the bipolar cells. Indeed, there is both anatomical and physiological evidence that all ganglion cells receive some amacrine cell input (see below). Nevertheless, the basic receptive field organization that these ganglion cells display appears to be derived from the bipolar cells.

The second type of ganglion cells found in the mudpuppy retina and other retinas, called the G_3 cell in Figure 4.17, responds transiently to retinal illumination, much as the transient amacrine cells do. This ganglion cell type in most species appears to receive much, if not all, of its synaptic input from the transient amacrine cells. Indeed, strong inhibitory (hyperpolarizing) input from the amacrine cells to these ganglion cells has been identified (Werblin, 1972; Miller, 1979), and transient, excitatory input to these cells has also been inferred in the carp and the rabbit* (Glickman et al.,

---

* In the mudpuppy there is evidence that the primary excitatory input to the on–off ganglion cells may not be from transient amacrine cells but from the bipolar cells (Miller and Dacheux, 1976d; but see Werblin, 1977a). In the rabbit and the carp acetylcholine appears to mediate much of the excitatory input to the on–off ganglion cells, and this agent strongly excites these cells in these species. In the mudpuppy acetylcholine has no effects on the on–off ganglion cells (Frumkes, 1981).

1982; Ariel and Daw, 1982a; Masland et al., 1984b). These on–off ganglion cells in many instances respond very well to motion, and they may show direction-sensitive responses (see Figure 2.14). This finding suggests, as does the comparative anatomy of the inner plexiform layer, that the amacrine cells are responsible for mediating complex ganglion cell activity such as motion and direction selectivity. Two kinds of experiments support this view. First, it has been shown that bipolar cells and other distal neurons do not usually show direction-sensitive responses (Werblin, 1970). Second, antagonists of known amacrine cell neurotransmitters abolish direction sensitivity in those cells displaying these properties in the rabbit retina (see Figure 5.7) and also abolish other complex properties of rabbit ganglion cells (Caldwell et al., 1978) (Chapter 5).

How amacrine cell interactions might account for direction-sensitive responses is yet to be determined. The transient amacrine cell responses, by being transient in nature and depolarizing at both the onset and cessation of illumination, are well suited for mediating the motion-sensing responses. For example, transient amacrine cells respond in a similar fashion to a bright spot on a dark background or dark spot on a light background, a feature of many motion-sensitive and direction-sensitive cells (Barlow and Levick, 1965; Michael, 1968b). But the situation is clearly more complex than that shown in Figure 4.17. For on–off, direction-sensitive ganglion cells in the rabbit, there appear to be at least two amacrine cells involved—one excitatory and one inhibitory. It seems likely, in view of the number of amacrine cells present in those retinas that have many direction-sensitive ganglion cells (see Chapter 3), that the formation of these receptive fields involves many amacrine cells.

Figure 4.18 shows two schemes that suggest how direction-sensitive responses may be mediated by excitatory and inhibitory transient amacrine cells (Dowling, 1970; Ariel and Daw, 1982b). Scheme a incorporates the main anatomical features of the retinas of those species demonstrating motion and direction sensitivity. For example, (1) the bipolar cells synapse mainly with the amacrine cells, (2) the ganglion cells receive their primary input from the amacrine cells, and (3) there are numerous serial synapses involved. In scheme a, the serial synapses are organized such that the first synapse (amacrine–amacrine) inhibits the second synapse (amacrine–ganglion cell). With the serial synapses arranged in the manner suggested in the figure, a spot moving from left to right (preferred direction) will evoke vigorous firing in the ganglion cell, whereas a spot moving from right to left (null direction) will evoke

4.18 Two schemes suggesting how direction-sensitive responses may be mediated by excitatory and inhibitory amacrine cells in the retina. In the top scheme, the inhibitory amacrine cell (A_I) makes its synapses on the excitatory amacrine cell (A_E) processes. In the lower scheme, the inhibitory amacrine cell makes its junctions directly on the ganglion cell (G) dendrites. Movement of a spot of light in the preferred direction activates initially the excitatory amacrine cell, thus causing vigorous firing of the ganglion cell in both schemes. Movement of a spot of light in the null direction activates first the inhibitory amacrine cell, thus causing inhibition of the excitatory amacrine cell (*top scheme*) or of the ganglion cell (*bottom scheme*) and a cancellation of the excitatory input to the ganglion cell. ○, excitatory synapses; ●, inhibitory synapses; B, bipolar cells providing input to the amacrine cells.

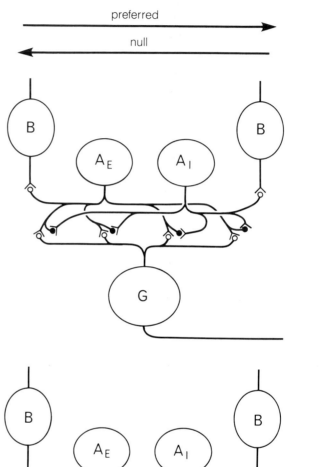

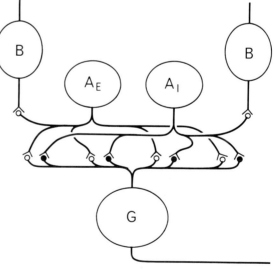

no firing in the ganglion cell. Scheme b is similar in most respects to a but involves direct inhibitory synapses on the direction-sensitive ganglion cells.

Within the last decade the neurotransmitters involved in the generation of on–off, direction-sensitive responses in the rabbit were identified. This discovery enabled investigators to test these two schemes. The excitatory input to these cells appears to be mediated mainly by acetylcholine (Ariel and Daw, 1982b; Masland et al., 1984b), whereas the inhibitory input is mediated principally by γ-aminobutyric acid (GABA) (Caldwell et al., 1978). GABA powerfully inhibits the release of acetylcholine from the rabbit retina (Massey and Redburn, 1982), a result that is in accord with scheme a. However, there is also evidence that GABA acts directly on ganglion cells (Ariel and Daw, 1982b); and both Werblin (1970) and Dacheux (Miller, 1979) have shown that movement in the null direction induces inhibitory postsynaptic potentials in direction-sensitive ganglion cells, in accord with scheme b. Thus, at present we are unable to decide which of the schemes shown in Figure 4.18 is more likely to be correct or whether the true situation is a combination of the two schemes. However, there seems little doubt that it is the transient amacrine cells that underlie the generation of these complex ganglion cell responses.

What is the role of the sustained amacrine cells in the formation of ganglion cell receptive fields? Studies carried out in the catfish retina are instructive in this regard. Three basic types of ganglion cells are found in this retina. The first type is a cell that has a small receptive field (200–300 μm) and gives sustained on- or off-responses to steps of light. Its receptive field is concentrically organized and is very bipolar-like (Davis and Naka, 1980; Lasater, 1982). This cell is very much like the G_1 cell of the mudpuppy retina (Figure 4.17). Cells of the second type have larger receptive fields—300–500 μm in diameter. These cells also usually respond with an on- or an off-response when their fields are illuminated with a centered spot of light. But their responses tend to be somewhat more transient, and a symmetrical antagonistic surround is usually not observed. Rather, there may be strong antagonistic areas adjacent to the central region of the receptive field, and often these antagonistic zones do not surround the central field (Davis and Naka, 1980). Characteristic of many of these cells is an orientation preference to bars or slits of light moved through the receptive field (Lasater, 1982). That is, the cells very often respond much more vigorously to a horizontal bar of light passing across the receptive field than to a vertically oriented slit of light, or vice versa.

In many respects, this second type of ganglion cells in the catfish resembles the orientation-sensitive cells in the rabbit retina that also do not demonstrate a symmetrical center–surround receptive field organization and that respond preferentially to oriented bars of light (Levick, 1967; Caldwell and Daw, 1978a). In the rabbit, infusion of GABA antagonists eliminates the orientation specificity of these cells, and the cells become conventional on- and off-center cells with a symmetrical surround (see Chapter 5). This finding suggests that these ganglion cells receive substantial inhibitory input from amacrine cells (input that underlies the orientation-specificity of the cell) in addition to bipolar input. In the catfish Naka (1977) has shown that the sustained amacrine cells, regardless of type (that is, whether depolarizing or hyperpolarizing), act to antagonize (that is, powerfully inhibit) the response of these ganglion cells to central illumination. That the sustained amacrine cells may underlie the more complex behavior of these ganglion cells is supported by the observation of Lasater (1982), who reported that GABA antagonists simplify the receptive field organization of these cells in the catfish.

How the sustained amacrine cells might impart orientation specificity to ganglion cells is suggested by the results of Naka (1980), who found that there is a class of sustained amacrine cells in the catfish whose dendritic fields are elliptical. The cells respond much more strongly to moving bars of light oriented with respect to the orientation of the dendritic field of the cell than to the bars of light oriented at right angles to the dendritic field orientation. Such cells, if inhibitory and superimposed on an on- or off-center ganglion cell, could very well give rise to the orientation preference shown by these cells. A similar mechanism has been suggested by Caldwell et al. (1978) for the receptive field organizations of the orientation-selective ganglion cells in the rabbit.

The third type of ganglion cells in the catfish is an on–off cell and appears to be very similar to the on–off cells observed in the mudpuppy and other animals (G_3 in Figure 4.17). It appears to receive its synaptic input predominantly from the transient amacrine cells (Chan and Naka, 1976). A simplified summary of the synaptic organization of the inner plexiform layer of the catfish retina is presented in Figure 4.19. The three ganglion cell types in the catfish are labeled G_1, G_2, and G_3, respectively. The G_1 cell receives its input predominantly from the bipolar cells, it has a dis-

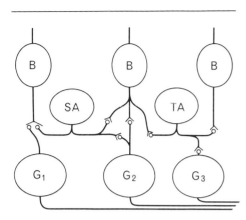

4.19 Wiring diagram for the three basic kinds of ganglion cells found in the catfish retina. The G_1 cell receives its input primarily from the bipolar cells (B). It is a small-field ganglion cell giving sustained on- and off-center responses to spots of light. The G_2 cell receives a mixed input from bipolar cells and sustained amacrine cells (SA). It is a large-field cell that gives more transient on- and off-center responses to spot illumination. Many of these cells appear to show orientation selectivity. The G_3 cell receives most of its input from the transient amacrine cells (TA). It is a large-field cell giving on–off responses to illumination presented anywhere in its receptive field. See text.

tinct center and surround receptive field organization, and its responses are sustained and bipolar-like. The G_2 cell receives both bipolar and sustained amacrine inputs. These cells respond to central illumination with on- and off-responses, but their receptive field organization is often complex and reflects substantial inhibitory input from the sustained amacrine cells. Finally, the G_3 cells receive their input predominately, if not exclusively, from the transient amacrine cells, and they respond to light with transient on–off responses.

A Summary Scheme

Figure 4.20 summarizes much of the information and many of the ideas concerning the functional organization of the vertebrate retina (Dowling, 1979). This diagram complements and adds to the highly simplified schemes shown in Figures 4.17 and 4.19, and it serves as a review of the major anatomical and physiological findings described in Chapters 3 and 4. The drawing again reflects mainly cone pathways, because less is known about the receptive field organization of rod pathways. In Figure 4.20, synapses are indicated as excitatory (open circles) or inhibitory (closed circles). In the retina, however, this is often not easy to determine, because many cells respond to light by hyperpolarizing. Thus, a synapse is indicated as excitatory if the postsynaptic response is of the same polarity as the response in the presynaptic element (an arrangement sometimes called a sign-conserving synapse) or if the postsynaptic response is enhanced as a result of synaptic interaction. Conversely, a synapse is indicated as inhibitory if the postsynaptic response is of opposite polarity (sign-inverting synapse) or if the postsynaptic response is diminished as a result of synaptic action. Known reciprocal interactions between two elements are also indicated (open triangles).

The receptor cells make chemical synapses with the bipolar and horizontal cells. The basal junctions made by the receptor cells onto the flat bipolar cells appear in many species to be excitatory. These flat bipolar cells (B_H) usually hyperpolarize in response to central field illumination, and their terminals end in the upper part of the inner plexiform layer. On the other hand, the ribbon synapses between receptor cells and invaginating bipolar cells act more like inhibitory synapses. These invaginating bipolar cells (B_D) usually depolarize in response to central field illumination, and their terminals usually end in the lower part of the inner plexiform layer.

4.20 Summary scheme of the synaptic interactions that occur in the retina and that underlie the receptive field properties of the on-center, off-center, and on–off ganglion cells. Excitatory synapses are indicated by open circles, inhibitory junctions by filled circles, and reciprocal junctions by open triangles. See text. Modified from Dowling (1979), with permission of MIT Press.

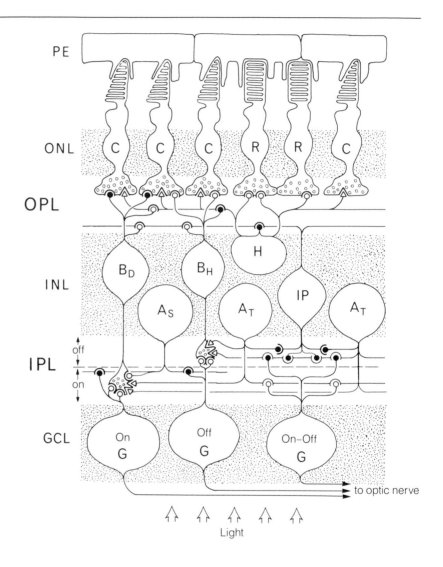

Receptor cells also drive the horizontal cells at the ribbon synapses, but these junctions appear to be excitatory, because the horizontal cells, like the receptors, generally hyperpolarize in response to illumination. Horizontal cells (H) often interact with the cone receptor terminals in a reciprocal fashion (open triangles); that is, the horizontal cells feed back onto the receptor cells, depolarizing them.

The horizontal cells also mediate the surround antagonism observed in bipolar cells, either presynaptically through the receptors or by direct interactions between horizontal and bipolar cells. The last chemical synaptic interactions occurring in the outer plexiform layer involve the interplexiform cells, about whose physiology we have little knowledge. In fishes, however, the interplexiform cells (IP in Figure 4.20) appear to depress the light-evoked activity of the horizontal cells, reduce the receptive field size of the horizontal cells, and enhance responsiveness of the bipolar cells (Hedden and Dowling, 1978; Negishi and Drujan, 1979; Mangel and Dowling, 1985). Thus, the interplexiform cells appear to modulate both the lateral inhibitory effects mediated by horizontal cells and the strength of center–surround interaction (see Chapter 5).

There are also electrical junctions observed in the outer plexiform and inner nuclear layers that are not shown in Figure 4.20. These junctions involve mainly the horizontal cells, and are between the cell perikarya in some species and between the cell processes in others. In all cases it appears that the electrical coupling between horizontal cells increases the receptive field size of the cells, thereby allowing the effects of the cells to be seen over a wide area.

In the inner plexiform layer the bipolar terminals appear to make primarily excitatory contacts with ganglion cells. The depolarizing bipolar cells contact the on-center ganglion cells, and the dendrites of these on-center ganglion cells extend mainly into the lower part of the inner plexiform layer. The hyperpolarizing bipolar cells contact the off-center ganglion cells and the dendrites of the off-center ganglion cells run predominately in the upper part of the inner plexiform layer.*

---

*A somewhat more complicated circuitry has been proposed to explain receptive field center responses of X-cells in the cat. On the basis of anatomical studies, McGuire et al. (1984, 1986) have suggested that pairs of on- *and* off-center bipolar cells feed into both on- and off-center ganglion cells in this species. One member of a bipolar cell pair is excitatory and the other inhibitory. So, for an on-center ganglion cell, the on-center bipolar is excitatory to the ganglion cells and the off-center bipolar inhibitory. Stimulation of the receptive field center causes a push–pull action; the excitatory on-center bipolar releases more transmitter, exciting the ganglion cell, whereas the inhibitory off-center bipolar releases less transmitter, also contributing to excitation of the cell. It has been proposed that the off-center ganglion cells operate in the opposite fashion; that is, the off-center bipolar is excitatory to the ganglion cell and the on-center bipolar inhibitory.

As yet there is no direct physiological evidence to support this hypothesis and some

As noted above, the sustained amacrine cells appear to interact principally with the on- and off-center ganglion cells, and they antagonize the response of these cells to central illumination; that is, they hyperpolarize the on-center ganglion cells and depolarize the off-center cells. Thus, the sustained amacrine cells provide an inhibitory input to these ganglion cells similar to the inhibition imparted by the horizontal cells onto the bipolar cells in the outer plexiform layer. The sustained amacrine cells, therefore, can contribute to the antagonistic surround response of the ganglion cells, as well as underlie more complex receptive field properties. Because all ganglion cells appear to receive some amacrine cell input, it seems likely that for most ganglion cells the antagonistic surround response will represent a combination of horizontal and sustained amacrine activity (see, for example, Caldwell and Daw, 1978b). If a ganglion cell receives relatively little amacrine cell input, its responses will strongly reflect the bipolar cell input, and it will give mainly sustained on- or off-responses and show a bipolar-like center and surround organization. On the other hand, if a ganglion cell receives substantial input from the sustained amacrines, its responses and receptive field organization will reflect properties of the sustained amacrines and perhaps show complex receptive field properties. This arrangement is illustrated by the responses of the large-field, on- and off-center ganglion cells in the catfish (Davis and Naka, 1980) and the orientation-sensitive ganglion cells of the rabbit (Caldwell et al., 1978).

On–off ganglion cells clearly receive most of their input from transient amacrine cells. In the rabbit the on–off, direction-sensitive ganglion cells are bistratified cells, receiving input in both the on- and off-laminae of the inner plexiform layer (Amthor et al., 1984). (Mechanisms for the generation of direction-sensitive responses of ganglion cells by the interactions of excitatory and inhibitory transient amacrine cells has already been discussed.)

The amacrine cells receive their input from the bipolar cells and other amacrine cells. Toyoda et al. (1973) and Miller (1979) have

---

pharmacological evidence argues against it. The glutamate analogue, 2-amino-4-phosphonobutyric acid (APB) blocks selectively all on-center bipolar responses (Slaughter and Miller, 1981, 1983a). The push–pull hypothesis predicts that APB, when applied to a retina, would affect both on- and off-center ganglion cells. In monkey lateral geniculate nucleus, where this has been looked at carefully, APB completely eliminates on-center cell activity but only transiently alters off-center cell responses (Schiller, 1984).

proposed that transient amacrine cells receive input from both the hyperpolarizing and depolarizing bipolar cells; the opposite polarities of the responses of the two cell types, coupled with small latency differences, could explain the generation of the on- and off-responses. We might suppose, therefore, that sustained amacrine cells receive their input from one or another of the bipolar cell types, depending on whether the sustained amacrine cell is hyperpolarizing or depolarizing. Reciprocal synapses between bipolar cells and amacrine cells have been described (shown in Figure 4.20 between the transient amacrine cells and the bipolar terminals). No evidence concerning the physiology of such interactions has yet been described. As noted previously, the feedback synapses between bipolar and amacrine cells may serve to enhance the transient nature of amacrine cell responses (Burkhardt, 1972).

The last interactions in the inner plexiform layer to be considered are those involving the interplexiform cells. Some evidence suggests that the interplexiform cells synapse mainly with the transient amacrine cells (Hedden and Dowling, 1978), but what role these neurons play in the inner plexiform layer is as yet unclear. By analogy to their role in the outer plexiform layer, they may modulate inhibitory interactions mediated by the transient amacrine cells. This notion that the interplexiform cells serve to modulate synaptic interactions in the retina will be discussed further in the next chapter.

In summary, the outer plexiform layer of the retina is concerned mainly with the static and *spatial* aspects of illumination. The neurons contributing processes to this layer respond primarily with sustained, graded potentials, and the neuronal interactions there accentuate contrast in the retinal image by forming an antagonistic center–surround organization at the level of the bipolar cells. The on- and off-center ganglion cells, receiving much of their input directly from either the center-depolarizing or center-hyperpolarizing bipolar cells, reflect this basic center–surround receptive field organization established in the outer plexiform layer. The inner plexiform layer, on the other hand, is concerned more with the dynamic or *temporal* aspects of photic stimuli. Transient amacrine and the on–off ganglion cells accentuate the changes in retinal illumination and respond vigorously to moving stimuli. Interactions in the inner plexiform layer underlie the motion- and direction-sensitive responses of the on–off ganglion cells, and the orientation-preferring responses of some of the on- and off-center cells.

Generation of Other Types of Ganglion Cell Receptive Fields

The foregoing discussion has focused on how the receptive fields of sustained on- and off-center ganglion cells, on–off direction-sensitive ganglion cells, and orientation-sensitive ganglion cells are formed. But, as pointed out in Chapter 2, a variety of other ganglion cell receptive field organizations have been described. How are these other receptive fields generated? As already noted, some of these receptive fields can be explained on the basis of different combinations of bipolar and amacrine cell inputs into a ganglion cell. An example is the Y-type ganglion cell in the cat, which shows a concentric center–surround receptive field organization but responses that are much more transient than are the responses of the sustained X-cells. An obvious explanation of their responses and receptive field properties is that these cells receive substantial input from the transient amacrine cells as well as from the bipolar cells. Evidence in favor of this explanation has been provided by Victor and Shapley (1979).

Some receptive field types can be explained on the basis of variation in the response properties of the input neurons. For example, the on, direction-sensitive ganglion cell can be thought of as a neuron that receives its input primarily from transient amacrine cells that respond only at the onset of illumination. Such amacrine cells have been observed; and except for the fact that the on, direction-sensitive units respond only at the onset of illumination, their properties are very similar to the on–off, direction-sensitive cells (Barlow et al., 1964; Caldwell et al., 1978).

The so-called edge detectors observed in rabbit, frog, and other retinas represent another kind of variation seen in on–off ganglion cells (Maturana et al., 1960; Levick, 1967). In the rabbit these cells are very similar in many of their properties to the direction-sensitive cells, but they are not sensitive to the direction of movement of the stimulus through the receptive field (Levick, 1967). These cells give on–off responses at all points within the receptive field, they respond as well to a dark spot as to a light spot, and they respond vigorously to moving stimuli. But they respond equally well to stimulus movement in all directions through the receptive field. It may be that the receptive fields of these ganglion cells are similar in their organization to those of the direction-sensitive cells, except that they receive no input from the amacrine cells that provide the

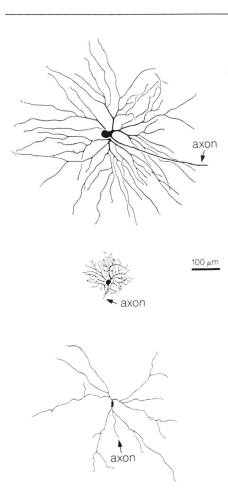

axon

100 μm

axon

axon

4.21 Three commonly observed types of ganglion cells in the periphery of the cat retina. At the top is an α or Y cell, in the middle a β or X cell, and at the bottom a γ or W cell. Modified from Boycott and Wässle (1974), with permission of the Physiological Society.

null-direction inhibition. This notion is supported by the observation that picrotoxin, which eliminates direction sensitivity in the direction-sensitive cells, has no effects on the local-edge detectors (Caldwell et al., 1978).

Other ganglion cell receptive fields are not explained so easily on the basis of known bipolar and amacrine inputs. Consider the so-called sluggish ganglion cells, which have been described in both cat and rabbit retinas (Cleland and Levick, 1974a; Caldwell and Daw, 1978a). The receptive fields of these cells are concentrically organized with on- and off-centers and antagonistic surrounds; some have sustained responses and others transient responses. Thus, in terms of their overall receptive field properties, they resemble the X- and Y-cells. The sluggish cells, however, have low spontaneous activity, their maximum firing frequency is lower than that for X- and Y-cells, and they respond only to very slowly moving stimuli. Furthermore, the axons of the sluggish cells conduct impulses very slowly. In the cat such ganglion cells have been identified anatomically by Boycott and Wässle (1974)—their γ-cells. These cells have large dendritic fields but small perikarya. The dendrites of γ-cells are thin all along their length, and there are considerably fewer dendrites on these cells than on the X- and Y-cells in the cat, the β and α cells, respectively, of Boycott and Wässle (1974) (Figure 4.21). An obvious question is whether these distinct morphological differences relate to any of the different physiological properties of these cells. In other words, could the sustained and transient sluggish cells have basically the same input as the X- and Y-cells but have somewhat different physiological properties because of their different cellular morphology (see Koch et al., 1982)? Although this is an interesting speculation, Ariel and Daw (1982a) have shown that the sluggish ganglion cells in the rabbit are affected by physostigmine, an acetylcholine antagonist, whereas the X- and Y-cells are not, a finding suggesting that the sluggish cells have an amacrine cell input not present on X- and Y-cells. Thus, it may be that the differences in physiological properties of these cells result both from a particular cellular morphology and from a somewhat different input.

FIVE

Synaptic Mechanisms and Chemistry

EVEN a partial understanding of the functional organization of the retina naturally leads to questions about retinal synaptic mechanisms. For example, it is well established that neurons release neurotransmitters when depolarized (Katz and Miledi, 1967). But when the retina is excited by light, most of the distal neurons hyperpolarize. How do the distal retinal synapses work? Can the responses of the cells postsynaptic to these hyperpolarizing cells be explained by a decrease in neurotransmitter release from the presynaptic cell, as was suggested in Chapter 4, and, if this is the case, why does the system work this way? Why do most of the distal retinal cells behave as if they are turned on in the dark and turned off in the light?

Other questions concern the substances used for synaptic transmission. Elsewhere in the brain, two classes of neuroactive agents are generally recognized as being released from terminals at chemical synapses, neurotransmitters and neuromodulators. Neurotransmitters are substances that act rapidly on postsynaptic cells, usually by directly altering membrane permeability to one or several ions and thus depolarizing or hyperpolarizing the cells. Neurotransmitters are responsible for mediating fast excitatory and inhibitory pathways in the brain. Neuromodulators, on the other hand, have been shown to act on postsynaptic cells by activating intracellular enzyme systems and affecting the functions of postsynaptic cells via biochemical mechanisms. Neuromodulators alter various aspects of neuronal function, and they appear to act over relatively long periods of time. They may not affect directly either membrane potential or membrane permeability.

A large number of substances are apparently released at synaptic sites in the retina. So far, at least fifteen substances have been identified as likely candidates for retinal neurotransmitters or neuromodulators (Table 5.1). Some of these, including L-glutamate, γ-aminobutyric acid (GABA), glycine, and acetylcholine, appear to act as classic neurotransmitters in the retina. Dopamine, on the other hand, appears to act as a neuromodulator, at least in the fish retina, where it has been extensively studied. The role of the other

Table 5.1 Neuroactive substances found in the retinas of several species

Amino Acids
　L-Aspartate
　γ-Aminobutyric acid (GABA)
　L-Glutamate
　Glycine

Amines
　Acetylcholine
　Dopamine
　Serotonin

Peptides
　Cholecystokinin
　Enkephalin
　Glucagon
　Neurotensin
　Neuropeptide Y
　Somatostatin
　Substance P
　Vasoactive intestinal peptide

neuroactive substances found in the retina, including at least eight neuropeptides, is still unclear. They may act as neurotransmitters, or as neuromodulators, or as both. It also must be noted that the list of substances released at retinal synapses is very likely to grow, because for a number of retinal cell subtypes no neurotransmitter or neuromodulator candidate has yet been found.

A final group of questions concerns the action of substances released by presynaptic terminals on the postsynaptic cells. Relatively little is known about the actions of various neuroactive substances on retinal cells, although such information is critical if we are to understand retinal mechanisms fully. In general, although we have learned much about retinal pharmacology over the past two decades, identifying with certainty the specific substances used at various synapses and discovering the mode of action of these agents has proved to be a difficult task. Thus, we can describe retinal anatomy and physiology in a more satisfactory way than we can retinal pharmacology—primarily because available pharmacological methods often give ambiguous and even misleading results.

Most pharmacological studies have employed whole eye or eyecup preparations or pieces of retina, and drugs have been applied by iontophoresis (ejecting a small amount of Ringer solution containing a drug from a pipette placed on or within the retina), by atomization (spraying a fine mist of drug onto the retinal surface), or by superfusion (bathing the retina in Ringer solution containing a drug). In the intact retina it is difficult, first of all, to assign observed drug effects to specific cells and to analyze rigorously the observed effects. Even when an effect is localized to a specific cell, the responses may have been altered by the activity of surrounding neurons or by the glial cells. Furthermore, in pharmacological experiments the concentrations of drugs applied, regardless of method, may not reflect the concentrations of drugs reaching the target.

To illustrate, it has long been known that the acidic amino acids—aspartate and glutamate—depolarize horizontal cells in many species, so these agents have been proposed as photoreceptor neurotransmitter candidates (Murakami et al., 1972). However, because horizontal cells are so extensively coupled electrically in vivo, it is virtually impossible to analyze the exact effects of these agents on the horizontal cells in the intact retina. Indeed in most studies investigators could not determine whether these agents opened channels, closed channels, or even acted directly on the horizontal cell membrane (see Nelson, 1973). Furthermore, horizontal cells in the intact retina are depolarized only with concentrations of aspartate and glutamate between 0.5 and 20 mM in the applied Ringer solution, concentrations thought to be beyond the physiological range. Because of this finding, many workers dismissed these agents as photoreceptor neurotransmitter candidates until recently, when it became clear that the intact retina contains powerful uptake mechanisms for the acidic amino acids, especially L-glutamate (Ishida and Fain, 1981; Marc and Lam, 1981). These uptake mechanisms alter the apparent sensitivity of the horizontal cells to L-glutamate by 25- to 200-fold (Ariel et al., 1984).

High concentrations of virtually any drug cause some effects on neurons. When drugs are applied by iontophoresis or atomization, accurate estimations of the amount of drug reaching a postsynaptic site usually cannot be made. Thus, the results of the experiments using these methods are often questionable. Even with superfusion and the use of low concentrations of highly specific transmitter antagonists, difficulties can be encountered. It has recently been shown in the carp retina that both bicuculline and picrotoxin—GABA antagonists—release dopamine from endogenous stores in the retina (O'Connor et al., 1986). Studies employing these antagonists may be confounded by the effects of dopamine on various cells, and the results may not simply reflect the direct blockade of GABA receptors.

Fortunately, a number of new approaches in retinal pharmacology have been introduced and promise to overcome a number of the problems of working with the intact retina. The use of receptor-specific analogues that are not taken up by neural tissue is one approach that has provided important information, and studying the pharmacology of isolated and cultured retinal cells is another. The latter approach appears particularly powerful for analyzing the precise mode of action of chemical agents on neurons because exact concentrations of drugs can be applied to localized regions of the cell (Tachibana, 1981; Lasater and Dowling, 1982; Ishida et al., 1984). Furthermore, it is relatively easy to measure conductance changes induced by neuroactive agents in isolated cells. Experiments with cultured cells have shown that horizontal cells are sensitive to concentrations of L-glutamate of less than 10 μM and that L-glutamate both opens and closes ionic channels in the horizontal cell membrane (Lasater and Dowling, 1982; Ariel et al., 1984; Tachibana, 1985; Kaneko and Tachibana, 1985b).

Distal Retinal Synaptic Mechanisms

The first clue that distal retinal neurons are stimulated by darkness and that light turns the cells off appeared in the first intracellular recordings of horizontal cell responses, which showed that the membrane potentials of horizontal cells in the dark are less negative (-25 to -40 mV) than those of other neurons (Svaetichin and MacNichol, 1958). Conversely, the membrane potential of L-type horizontal cells hyperpolarizes in the light to a level that is similar to the resting potential of most neurons (-60 to -80 mV). Thus, horizontal cells appear to be maintained in a partially depolarized state in the dark, and light decreases this depolarization. Trifonov and Byzov (1965) later showed that currents that are passed through the retina and that depolarize the receptor terminals caused a depolarization of the horizontal cell, and currents that hyperpolarize the receptor terminals hyperpolarized the horizontal cells (see also Byzov and Trifonov, 1968). On the basis of these experiments, Trifonov (1968) proposed that photoreceptors continuously release a depolarizing transmitter in the dark and that light decreases the flow of this transmitter. Subsequent studies supported this notion by showing that there is a steady inward flow of Na^+ into the outer segments of the photoreceptors in the dark and that light decreases the Na^+ conductance of the outer segment, thereby causing the cells to hyperpolarize (Hagins et al., 1970). In other words, the photoreceptors are partially depolarized in the

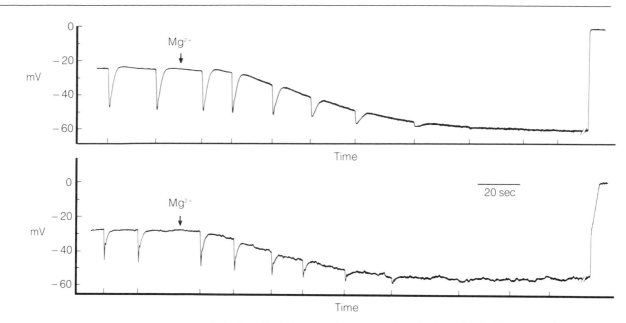

5.1   The effect of magnesium on two skate horizontal cells. The marks along the time scales indicate light (flash) presentations. Ringer solution containing magnesium was applied to the retina (*arrows*), and within 15–25 sec the cells began to hyperpolarize. Over the next few minutes the cells hyperpolarized to approximately −60 mV and light-evoked activity was lost. At the end of the trace, the pipettes were withdrawn from the cells. The large, fast, positive shifts of potential upon pipette withdrawal confirmed that the resting potentials had increased to −55 to −60 mV following Mg²⁺ application. From Dowling and Ripps (1973), reprinted by permission from *Nature* 242:101–103, copyright © 1973, Macmillan Journals Ltd.

dark, a condition consistent with the view that transmitter is being released from receptor synapses in the dark.

An experiment that tests this hypothesis is shown in Figure 5.1 (Dowling and Ripps, 1973). It is well known that high levels of extracellular Mg^{2+} and certain other divalent cations such as Co^{2+} and Mn^{2+} block neurotransmitter release from the presynaptic terminals at most chemical synapses (Del Castillo and Katz, 1954; Takeuchi and Takeuchi, 1962; Katz and Miledi, 1967). If photoreceptors release transmitter in the dark, application of Mg^{2+} to the retina should block this release and cause horizontal cells to hyperpolarize as they do in the light. Figure 5.1 shows that within a few minutes after the application of Mg^{2+} to skate retinas, horizontal cells hyperpolarized from the dark potential of about −30 to −60 mV and light responses were lost.

All of the available evidence supports the idea that in the dark photoreceptors release a transmitter that depolarizes horizontal cells and that light decreases the release of this substance (Ripps et al., 1976; Waloga and Pak, 1976). Can bipolar cell activity be explained on a similar basis?

There are two basic physiological types of bipolar cells in the vertebrate retina: one that hyperpolarizes in response to illumination of the receptors in the receptive field center and another that depolarizes. The hyperpolarizing (off-center) bipolar cells exhibit an increase in membrane resistance during light stimulation of the center of their receptive fields (Toyoda, 1973; Nelson, 1973); thus, generation of their light responses can be explained by the withdrawal of a transmitter that open channels in the bipolar cell membrane (Saito and Kaneko, 1983). The responses of the depolarizing (on-center) bipolar cells, however, are more difficult to understand. In the light the center response of these cells in the mudpuppy retina (Nelson, 1973) and the rod bipolar cells in the carp retina (Saito et al., 1979) are accompanied by a *decrease* in membrane resistance. If we assume that these bipolar cells receive their central input directly from the receptors, then the effect of the receptor neurotransmitter must be to decrease the conductance of the bipolar cell membrane to one (Na^+, for example) or several ions. Thus, in the light, when transmitter release is decreased, a conductance increase would be observed in the bipolar cell. Although decreasing membrane conductance is an unconventional action for a neurotransmitter, it has been shown to occur elsewhere in nervous systems (see, for example, Weight, 1974).

Not all on-center bipolar cells appear to work this way. The center responses of the on-center, cone-dominated bipolar cells in the carp show an *increase* of membrane resistance in the light (Saito et al., 1979). This finding suggests that the transmitter, which is released from the receptors in the dark, opens channels (K^+ or Cl^- channels or both) in the membranes of these bipolar cells and that in the light these channels close, thereby causing the cell to depolarize.

Evidence that off-center bipolar cells are maintained in a depolarized state in the dark and that on-center bipolar cells are maintained in a hyperpolarized state has been provided by Dacheux and Miller (1976), who showed that 2 mM Co^{2+}, which blocks synaptic transmission in the mudpuppy retina, hyperpolarizes the off-center bipolar cells and depolarizes the on-center bipolar cells. Furthermore, Murakami et al. (1975) have shown that glutamate and aspartate—photoreceptor transmitter candidates—depolarize off-center bipolar cells and hyperpolarize on-center bipolar cells in the carp retina. These findings indicate, as suggested in Chapter 4, that the synapse between the receptors and the hyperpolarizing bipolars can be viewed as excitatory, whereas the synapse between the receptors and the depolarizing bipolar cells is analogous to an inhibitory junction.* In addition, as noted in Chapter 3, the basal junctions made by the receptors onto the dendrites of the flat bipolars have morphological characteristics consistent with those of excitatory synapses, whereas the junctions between receptors and invaginating bipolars appear inhibitory in nature. Thus, single vertebrate photoreceptors appear to make both excitatory-like and inhibitory-like synapses on postsynaptic elements. This situation is highly unusual because most neurons in the vertebrate brain make either all excitatory or all inhibitory junctions with postsynaptic cells and all of their synapses look the same. But, although these two kinds of receptor junctions appear to have different morphologies, they do seem to use the same neurotransmitter (see below).

We assume that other retinal neurons that are maintained in a partially depolarized state in the dark—namely, the horizontal and off-center bipolar cells—also release transmitters in the dark and that their light-evoked effects are exerted by the withdrawal of transmitter. As yet, little is known about the synaptic effects exerted by the horizontal and the off-center bipolar cells, but what has been reported is consistent with this assumption. Off-center ganglion cells of the mudpuppy were hyperpolarized when 2 mM Co^{2+} was perfused onto a retinal eyecup preparation (Figure 5.2). Their light-evoked activity was virtually abolished and the hyperpolarization induced by Co^{2+} was associated with an increase in membrane resistance (Miller, 1979). The results of this experiment suggest that the off-center bipolar cells in the mudpuppy release an excitatory (depolarizing) transmitter in the dark and that light decreases the release of this substance. Similar experiments with on-center ganglion cells, on the other hard, showed neither a hyperpolarization nor a change in input resistance during Co^{2+} application. This result suggests that the on-center bipolar cells release little neuro-

---

* The synaptic mechanism underlying the responses of the on-center bipolar cells that show a *decrease* in membrane resistance in the light is quite different from classic inhibition (see Appendix). More properly, it might be termed disfacilitation, to recognize the fact that the transmitter released from the photoreceptors closes channels in the bipolar cell membrane.

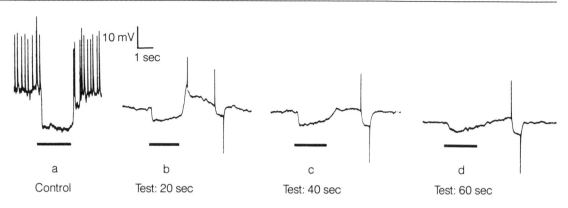

5.2 Light-evoked responses of an off-center ganglion cell in the mudpuppy retina before (**a**, control) and after (**b**, at 20 sec; **c**, at 40 sec; **d**, at 60 sec) the application of 2 mM Co^{2+}. The duration of the light stimulus is indicated by the heavy bars below the traces. Within 20 sec of Co^{2+} application, the cell hyperpolarized below spike-firing threshold and the light-evoked response was reduced in amplitude. At the end of each of the three traces recorded in the presence of Co^{2+}, a brief, negative, 0.1-nA current pulse was delivered to the cell. The increase in amplitude of the voltage deflection recorded with time indicates the Co^{2+} induced an increase of membrane resistance in parallel with the hyperpolarization of the cell. Modified from Miller (1979), with permission of MIT Press.

transmitter in the dark. Experiments indicating that horizontal cells also release neurotransmitter in the dark will be described later.

Another important question with regard to synaptic mechanisms in the distal retina concerns the amount of voltage change across the presynaptic membrane that is required to alter transmitter flow and to enable the postsynaptic cell to detect a signal. Receptor and bipolar cells respond to light with sustained, graded potentials whose maximum amplitudes are only 20–30 mV. Furthermore, with dim illumination, the responses of these cells will only be a small fraction of their maximum amplitude. At other chemical synapses that have been studied, potential changes across the presynaptic membrane of more than 20–30 mV are required for any significant release of neurotransmitter to occur (Takeuchi and Takeuchi, 1962; Katz and Miledi, 1967; Kusano, 1970).

Measurements on toad rods stimulated with near-threshold light have provided some interesting insights into this problem (Fain et al., 1976). It has long been known that rods respond to a single quantum of light and that the perception of light requires the absorption of one quantum by each of 5 to 10 rods in a field of approximately 5,000 rods (Hecht et al., 1942; Pirenne, 1962). When a field of toad rods is illuminated with a light providing an average of only one quantum per receptor, a voltage change of about 1 mV is generated in each rod (see Chapter 4). But when the light is dimmed to near-threshold levels (that is, one quantum absorbed per 500–1,000 rods), all rods continue to register a voltage change because of the electrical coupling between photoreceptors. The response generated in any one rod, however, including those that capture a quantum, is reduced significantly because of the coupling. Indeed, under threshold conditions, the voltage changes in all rods are so small that they cannot be detected with present techniques. A calculation indicates that a potential change of no more than 50–100 μV occurs in any one rod at threshold illuminance levels (Fain, 1977). A similar calculation for turtle cones suggested that at threshold the receptor potential change was only 5–10 μV in amplitude (Fain et al., 1977). These results indicate, therefore, that exceedingly small presynaptic voltage changes are capable of modulating flow of synaptic transmitter at the receptor terminal. It may well be that synapses made by other retinal neurons that respond with graded potentials are also highly sensitive to voltage changes.

How the photoreceptor synapse manages this high degree of voltage sensitivity is not known. This phenomenon may relate to the unusual nature of the photoreceptor synapse. It is noteworthy that both hair cells in the ear and electroreceptors also appear to alter transmitter flow in response to very small changes in presynaptic voltage (Harris et al., 1970; Bennett, 1970), and both of these receptor cells, like the photoreceptors, make ribbon synapses. Furthermore, that photoreceptor cells are maintained in a partially depolarized state in the dark provides another partial explanation for their high sensitivity. The relation between presynaptic voltage and transmitter release is an S-shaped function. As noted above, transmitter release begins when the terminal is depolarized by 20–30 mV and is maximal when the terminal is depolarized by 50 mV or more from rest (Katz and Miledi, 1967). If the photoreceptor has a similar voltage–release relationship, it would be well along this function in the dark (that is, on the steep part of the curve), because the rod is depolarized by 30–40 mV in the dark. Thus, small, light-induced changes in the presynaptic voltage will significantly alter transmitter flow (Figure 5.3). This consideration may

5.3 Input–output relationship for a giant synapse in the squid. Pre- (V_{pre}) and postsynaptic (V_{post}) voltages are given with respect to their resting membrane potentials. Because photoreceptors are highly depolarized in the dark, they presumably operate along the steep part of the curve (*arrows*). From Fain (1977; reprinted with permission of Academic Press), who redrew the data from Katz and Miledi (1967).

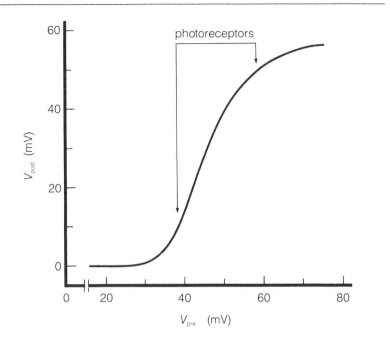

explain why receptors and other retinal neurons are maintained in a partially depolarized state in the dark.

Pharmacology of Retinal Synapses

Receptors and L-Glutamate

As already noted, it has long been known that L-glutamate and L-aspartate depolarize horizontal cells (Figure 5.4), and thus they have long been considered as candidates for photoreceptor neurotransmitters (Murakami et al., 1972; Dowling and Ripps, 1972; Cervetto and MacNichol, 1972). Depolarization of horizontal cells by these agents has been observed in all retinas so far examined. But in most studies employing these agents, it has not been possible to discriminate between the effects of these substances; and usually no effects of these agents are seen in intact retinas below concentrations of 0.5 to 20 mM, levels far beyond the concentrations expected physiologically (Ariel et al., 1984; Bloomfield and Dowling, 1985a). Nevertheless, recent studies have provided rather compelling evidence that L-glutamate, or a glutamate-like molecule, is likely to be employed as a neurotransmitter by many photorecep-

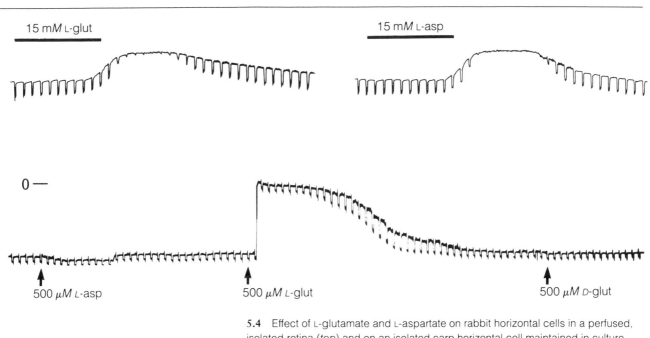

5.4 Effect of L-glutamate and L-aspartate on rabbit horizontal cells in a perfused, isolated retina (*top*) and on an isolated carp horizontal cell maintained in culture (*bottom*). In the intact retina, both L-glutamate and L-aspartate depolarize horizontal cells and abolish light-evoked responses (downward deflections); but high (millimolar) concentrations of these agents are required to obtain these effects. On isolated horizontal cells, L-glutamate is effective at much lower (micromolar) concentrations, whereas L-aspartate and D-glutamate are ineffective. The downward deflections in this record reflect the responses of the cell to constant current pulses injected into the cells to estimate membrane resistance (see legend of Figure 5.2). Note that immediately after the cell depolarized following the application of L-glutamate, the size of the responses to the current pulses were virtually the same as before the drug application. This indicates that the potential change occurred without a significant change in membrane resistance. *Top,* modified from Bloomfield and Dowling (1985a), with permission of the American Physiological Society; *bottom,* from Lasater and Dowling (1982), reprinted with permission.

tors, both rods and cones, in many species. First, it has been shown that horizontal and bipolar cells in the intact retinas of a variety of animals are exquisitely sensitive to the specific glutamate analogues kainate and quisqualate and that low micromolar concentrations of these agents powerfully depolarize the cells (Sheills et al., 1981; Slaughter and Miller, 1983c; Ariel et al., 1984; Bloomfield and Dowling, 1985a). On the other hand, horizontal cells in situ are not depolarized by even rather high concentrations of the specific aspartate analogue N-methyl-D-aspartate (NMDA). Indeed,

NMDA often acts like an antagonist on horizontal cells (Rowe and Ruddock, 1982; Ariel et al., 1984). Second, isolated and cultured horizontal cells respond to low (<10 μM) concentrations of L-glutamate and usually not at all to L-aspartate (Figure 5.4b), (Lasater and Dowling, 1982; Ariel et al., 1984; Ishida et al., 1984). Furthermore, the isolated horizontal cells are strongly depolarized by low micromolar concentrations of quisqualate and kainate but not by NMDA or a variety of other retinal transmitter candidates. Indeed, NMDA usually slightly hyperpolarizes isolated horizontal cells (Ariel et al., 1984).

Experiments on isolated cells have also produced insights into the mechanism of action of L-glutamate on horizontal cells. Interestingly, the effects of L-glutamate on isolated horizontal cells are surprisingly complex. Some hint of this can be gleaned from the lower record in Figure 5.4, which shows that little overall membrane resistance change occurred following L-glutamate application to an isolated horizontal cell, even though glutamate induced a membrane potential change of nearly 80 mV in the cell. Such a result suggests that L-glutamate both opens and closes channels in the horizontal cell membrane (Nelson, 1973); that is, L-glutamate appears to open channels in the cell membrane that allows mainly Na^+ to flow into the cell, and, at the same time, to close K^+ channels (Lasater and Dowling, 1982; Tachibana, 1985; Kaneko and Tachibana, 1985b). Both effects lead to a depolarization of the isolated cells, which, if large enough, results in a prolonged Ca^{2+}-dependent action potential (Figure 5.5). Thus, following L-glutamate application to isolated horizontal cells, at least three different channels in the cell membrane may be altered: glutamate-sensitive Na^+ and K^+ channels and voltage-sensitive Ca^{2+} channels.

Following repetitive stimulation of the presynaptic terminal, desensitization of the postsynaptic cell is observed at many synapses; that is, the postsynaptic membrane becomes significantly less responsive to applied neurotransmitter over time (Katz and Thesleff, 1957). Vertebrate photoreceptors release neurotransmitter continuously in the dark; and an important question is whether desensitization of the postsynaptic membrane occurs at these junctions. A partial answer comes from experiments in which L-glutamate was repetitively or continuously applied onto isolated horizontal cells of the skate and the goldfish. No desensitization of the postsynaptic receptors on these cells occurred following application periods of up to 15 minutes (Lasater et al., 1984; Ishida et al., 1984).

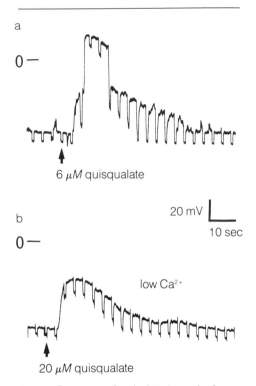

5.5 a Response of an isolated carp horizontal cell to a low concentration (6 μM) of quisqualate, an L-glutamate analogue. Initially the drug induced a slow depolarization, which was accompanied by a membrane resistance increase (as shown by the increase in amplitude of the current pulse responses). Superimposed on this slow depolarization was a rapid regenerative response that was accompanied by a membrane resistance decrease (that is, the responses to the current pulses decreased in amplitude).

b The response of an isolated horizontal cell to 20 μM quisqualate in Ringer solution lacking Ca^{2+}. Only a slow depolarizing response was evoked in the cell, a result indicating that the regenerative component evoked in normal Ringer solution (*top trace*) is likely to depend on Ca^{2+}. Modified from Lasater and Dowling (1982), with permission.

The pharmacological properties of the acidic amino acid receptors on both horizontal and bipolar cells in the intact mudpuppy retina have been probed using the glutamate analogue 2-amino-4-phosphonobutyric acid (APB) and the acid amino acid antagonist (±)*cis*-2,3-piperidinedicarboxylic acid (PDA) (Slaughter and Miller, 1981, 1983a). APB was found to block selectively the on-center bipolar response while leaving the off-center bipolar and horizontal cell responses intact. Conversely, PDA blocked the off-center bipolar and horizontal cell responses while leaving the on-center bipolar cell responses intact. These results indicate that the acidic amino acid receptors on the on-center bipolar cells are distinct from the acidic amino acid receptors on the horizontal and off-center bipolar cells. Other experiments have shown that the effect of the photoreceptor transmitter on many on-center bipolar cells is to close ionic channels, whereas its action on the off-center bipolar cells and the horizontal cells is to open channels. Thus, a difference in the postsynaptic receptors on these cells is to be expected. Both types of receptors, however, are strongly activated by L-glutamate and the glutamate analogues kainate and quisqualate. Hence, both act as glutamate-preferring receptors.

In the mudpuppy retina the glutamate analogue O-phospho-D-serine (DOS) distinguishes the glutamate-preferring receptors found on horizontal cells from those on off-center bipolar cells (Slaughter and Miller, 1985). That is, DOS antagonizes horizontal cell responses but has little effect on the center response of on-center and off-center bipolars. These data indicate that in the mudpuppy the glutamate-preferring receptors on all three types of second-order neurons receiving input from the photoreceptors may be distinguished pharmacologically and are likely to be somewhat different molecules.

Finally, several investigators have suggested that acetylcholine may be a photoreceptor neurotransmitter, particularly in certain cold-blooded vertebrates (Lam, 1972; Gerschenfeld and Piccolino, 1979). None of the evidence presented in support of this notion is compelling, and there is some negative evidence (Masland and Mills, 1979; Dowling et al., 1983). Thus, most investigators agree that the neurotransmitter used by most rods and cones is L-glutamate or a glutamate-like substance. There is some evidence, however, that blue cones in the carp retina may use a different transmitter (Mangel et al., 1985), so other photoreceptor transmitters may be found.

Horizontal Cells and GABA

In most cold-blooded vertebrates and birds many horizontal cells avidly take up exogenously applied GABA, a finding suggesting that this inhibitory neurotransmitter agent is released by certain of the horizontal cells in these animals (Lam and Steinman, 1971; Voaden, 1976). In elasmobranch fishes and mammals horizontal cells generally do not take up GABA, and the identity of the horizontal cell transmitter in these animals is unknown (Lam, 1975b). In teleost fishes and turtles the evidence in favor of GABA being a horizontal cell neurotransmitter is compelling. Synthesis and release of GABA from isolated goldfish horizontal cells has been reported (Lam, 1975a; Ayoub and Lam, 1984), and GABA has been shown to have postsynaptic effects appropriate for the horizontal cell neurotransmitter (Lam et al., 1978; Tachibana and Kaneko, 1984). However, in both the goldfish and the turtle, it has been shown that GABA is likely to be the neurotransmitter of only one subtype of horizontal cells, namely, the cone-related L-type (H1) cells (Marc et al., 1978). What the transmitter of the other subtypes of horizontal cells may be is unknown.

Physiological evidence indicates that horizontal cells provide negative feedback onto cones and that horizontal cells are depolarized in the dark (Chapter 4). Thus, it is likely that in the dark horizontal cells continuously release GABA, which hyperpolarizes cones. In the light, as the horizontal cells hyperpolarize, transmitter flow decreases and the receptors depolarize. In support of this view, a continuous release of GABA from goldfish retinas maintained in the dark has been found (Marc et al., 1978; Ayoub and Lam, 1984; Yazulla, 1985). And Baylor et al. (1971) have shown that hyperpolarizing currents passed into horizontal cells cause a depolarization of turtle cones. Finally, Tachibana and Kaneko (1984) have demonstrated, using isolated turtle cones, that GABA opens Cl⁻ channels in red- and green-sensitive cones, which in the intact retina would hyperpolarize the cells. Interestingly, neither blue-sensitive cones nor rods in the turtle appear to be sensitive to GABA.

The release of GABA from horizontal cells has been studied both in intact retinas and in isolated horizontal cells. In the intact retina both L-glutamate and L-aspartate are effective in releasing GABA, but only at millimolar concentrations (Yazulla and Kleinschmidt, 1983). The glutamate analogues quisqualate and kainate are also effective in releasing GABA from intact retinas, at micromolar concentrations (Ayoub and Lam, 1984; Cha et al., 1986). In isolated horizontal cells L-glutamate is considerably more potent than L-aspartate in releasing GABA, and it is effective at low micromolar concentrations (Ayoub and Lam, 1984). Potassium also readily releases GABA from horizontal cells in both kinds of preparations.

Curiously, most of the release of GABA from horizontal cells appears to be Ca^{2+} independent (Schwartz, 1982; Yazulla and Kleinschmidt, 1983; Ayoub and Lam, 1984), and it has been suggested that GABA is not usually released from horizontal cell processes via synaptic vesicles, as is the case at most synapses. It has been proposed, rather, that much of the release of GABA is by a carrier or transport mechanism, perhaps the same mechanism that accounts for the substantial uptake of GABA into the horizontal cells, but operating in reverse (Schwartz, 1982). It has also been suggested that the failure to observe typical synapses made by the horizontal cells (back onto receptor terminals, for example) relates to the unconventional release of much of the GABA from horizontal cells and their processes (Schwartz, 1982; Yazulla and Kleinschmidt, 1983).

Bipolar Cells

As yet, there is very little information available concerning transmitters employed by bipolar cells. As noted earlier, physiological evidence indicates that the bipolar cells release an excitatory neurotransmitter (Naka, 1976; Miller and Dacheux, 1976c). In the mudpuppy an excitatory amino acid antagonist, (±)cis-2,3-piperidinedicarboxylic acid (PDA), blocks on-center ganglion cell activity but not on-center bipolar cell responses (Slaughter and Miller, 1983b). This result indicates that the antagonist is specifically blocking the transmitter released by the on-center bipolar cells, and it suggests that the on-center bipolar cell transmitter is an acidic amino acid. PDA also blocks the responses of off-center ganglion cells, but it also blocks off-center bipolar cells, thereby making it impossible to determine the site of action of the drug in this case. In the rabbit retina glutamate and aspartate and their analogues quisqualate, kainate, and NMDA all have excitatory effects on both amacrine and ganglion cells; and evidence was obtained that these agents were acting directly on the inner plexiform layer neurons (Bloomfield and Dowling, 1985b). These experiments are consistent with the notion that bipolar cells release an acidic amino acid (see also Ikeda and Sheardown, 1982). On the other hand, these experiments do not exclude the possibility that glutamate or aspartate or both are released in the inner plexiform layer by amacrine cells.

Uptake studies have indicated that some bipolar cells in some species will take up transmitter precursors or the neurotransmitters themselves. For example, some bipolar cells in the chicken accumulate choline (Baughman and Bader, 1977), some bipolar cells in the cat accumulate glycine (Pourcho, 1980), and some bipolar cells in the skate take up serotonin (Brunken et al., 1986; B. Ehinger, personal communication). On the basis of these studies, it has been proposed that these substances may serve as bipolar cell transmitters. Others question the significance of these observations (Cohen and Sterling, 1986); no physiological evidence that these agents are bipolar cell transmitters has been provided.

Amacrine Cells

The number of transmitters associated with amacrine cells is surprisingly large. In several retinas five well-established neuroactive substances (GABA, glycine, acetylcholine, dopamine, and serotonin) and eight neuropeptides (substance P, enkephalin, neurotensin, somatostatin, neuropeptide Y, vasoactive intestinal peptide, cholecystokinin, and glucagon) have been found in amacrine cells. In many cases a morphological type of amacrine cells has been correlated with a specific neurotransmitter substance. Thus, there appear to be pharmacological types of amacrine cells that contain and use different substances. Furthermore, it appears that these pharmacological types of amacrine cells make specific and different connections within the inner plexiform layer (see, for example, Figure 5.9) and mediate different functions. Some neurotransmitters appear to be present in more than one morphological type of amacrine cells; and there is evidence (to be described later) that two or perhaps even more neuroactive substances may be present in the same amacrine cell. Some of the pharmacological types appear to be rare, making up less than 1 percent of the amacrine cell population. But other types appear to be very common, particularly the GABA- and glycine-accumulating amacrine cells (Pourcho, 1980). In the cat at least four morphological types of amacrine cells take up GABA, and these GABA-accumulating amacrine cells account for 38 percent of the total amacrine cell population. Glycine is taken up by three morphological types of amacrine cells in the cat, and these three types are distinct from the four types that take up GABA. The glycine-accumulating cells account for an additional

43 percent of the amacrine cell population in the cat, so over 80 percent of the amacrine cells in the cat take up either GABA or glycine. Whether all of these cells release GABA or glycine is not known, but it is believed that most of them do.

There is abundant evidence that both GABA and glycine are inhibitory transmitters in the inner plexiform layer, and amacrine cells containing these substances are very numerous. So it is not surprising that inhibitory amacrine cell activity is readily observed physiologically (Miller et al., 1981a,b). Both transient and sustained inhibition has been observed in ganglion cells, presumably mediated by transient and sustained amacrine cells that employ either GABA or glycine as their transmitter. Evidence for excitatory input from amacrine cells is also strong, particularly from cholinergic (acetylcholine-releasing) amacrine cells onto on–off ganglion cells (Ariel and Daw, 1982b; Glickman et al., 1982; Masland et al., 1984b). About the physiological effects of most other amacrine cell transmitters, however, very little is known.

GABA AND GLYCINE   As already noted, GABA and glycine appear to be the most common neurotransmitters used by amacrine cells in the inner plexiform layer. They both have effects on virtually all ganglion cells in the mudpuppy (Miller et al., 1981b), and the same probably holds true for other species. Figure 5.6 shows the effects of GABA and glycine on a mudpuppy off-center ganglion cell before and after the addition of Co^{2+} to the bathing solution (Ringer). Initially, the cell was strongly hyperpolarized by both agents; and there was a large drop in membrane resistance during the hyperpolarizing response, a result indicating that these transmitters were opening channels in the cell membrane. The addition of Co^{2+} to the bathing medium, which blocked synaptic input, caused the cell to hyperpolarize (see Figure 5.6). However, both GABA and glycine still had striking effects. This finding indicates that these substances were acting directly on the ganglion cell membrane; and other experiments suggested that the effects of GABA and glycine in the mudpuppy are mediated, at least in part, by alterations in Cl^- permeability of the membrane. The addition of the GABA and glycine antagonists (picrotoxin, bicuculline, and strychnine) to the bathing medium enhances the light responses of the ganglion cells, a result suggesting that in the inner retina there is a continuous (dark) release of GABA and glycine. Light-evoked inhibitory effects appear to be superimposed on a small continuous (tonic) GABA and glycine inhibition. There is also biochemical evidence for tonic

5.6 Effects of glycine and GABA on an off-center ganglion cell in the mudpuppy. Both agents hyperpolarized the cell and abolished spike-firing (*top trace*). The response of the cell to these agents was accompanied by a significant decrease in membrane resistance (that is, the responses to the negative current pulses are in a positive [upward] direction during drug application). When exposed to Co^{2+}-containing Ringer (*bottom trace*), the cell hyperpolarized by 10 mV and spike-firing was lost. The cells still responded to both glycine and GABA, however, a result indicating that the effects of the agents are direct ones. The dashes below each trace represent light flashes. RMP, resting membrane potential. From Miller et al. (1981b), reprinted with permission of the American Physiological Society.

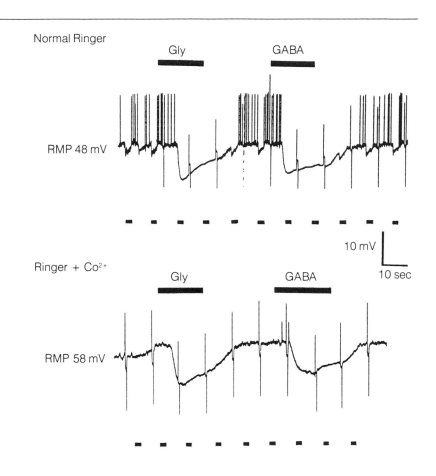

GABA release in the inner plexiform layer of both rabbit and carp retinas.

The effects of GABA and glycine on ganglion cell receptive field organization have been studied carefully in the rabbit retina using the antagonists picrotoxin, bicuculline, and strychnine (Caldwell and Daw, 1978b; Caldwell et al., 1978). As pointed out in Chapter 4, these agents change specific ganglion cell receptive field properties, particularly those of the complexly organized receptive fields. The infusion of picrotoxin or bicuculline abolishes direction sensitivity in both the on–off and the on-center, direction-sensitive ganglion cells and eliminates orientation selectivity in the orientation-sensitive ganglion cells.

Figure 5.7 shows the effects of picrotoxin on an on–off, direction-sensitive unit. The cell was activated with a stimulus large enough to elicit responses at both its leading and trailing edges as

5.7 Responses to picrotoxin in an on–off, direction-sensitive ganglion cell from the rabbit. This GABA antagonist abolished direction sensitivity for both moving light (*top*) and dark (*bottom*) spots. Responses in the preferred direction of movement are shown on the right; responses in the null direction on the left. Each traverse of the receptive field by the large spots resulted in a double burst, one corresponding to the leading edge of the spot, the other to the trailing edge. Following picrotoxin infusion, the spots moving in the null direction elicited strong responses. Modified from Caldwell et al. (1978), with permission of the Physiological Society.

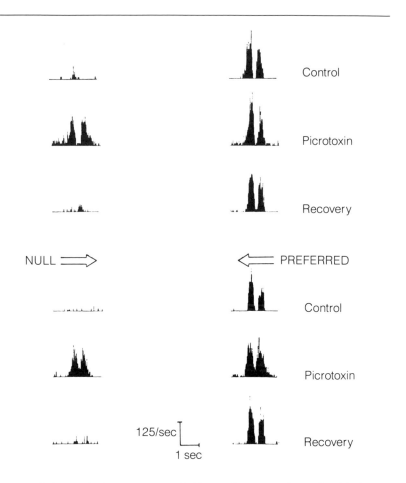

it passed through the receptive field (explaining why a double burst of activity was evoked by stimuli in the preferred direction). The records on the right are for the preferred direction, on the left for the null direction. Initially, no convincing response was seen when the stimulus was passed across the field in the null direction; the few spikes observed represent spontaneous, on-going activity, which is observed in many ganglion cells, especially in mammals. Following the infusion of picrotoxin into the animal, the response to movement in the null direction was nearly as great as that induced by movement in the preferred direction. In other words, the direction-sensitive responses of this cell were essentially abolished by picrotoxin, while the responses to movement in the preferred direction were slightly enhanced but otherwise unaffected. Following wash-out of the antagonist, strong direction sensitivity returned to the cell.

Orientation-specific ganglion cells in the retina of the rabbit respond preferentially to moving or stationary bars of light in either the vertical or horizontal plane, and they may respond either at the onset or offset of illumination. Following the infusion of picrotoxin (or bicuculline) into the retina, the orientation selectivity of the cells is lost and the receptive fields typically become concentric and show a center–surround antagonism. It would appear, then, that orientation selectivity is imposed on an on- or off-center ganglion cell by inhibitory input onto the cell, mediated by GABAergic (GABA-releasing) amacrine cells (see Chapter 4).

The infusion of strychnine into the rabbit caused somewhat subtler effects on the properties of the complex ganglion cell receptive fields. Many of the movement-sensitive cells in rabbit normally respond to a restricted range of velocities. Infusion of strychnine often broadened the range of velocities to which a particular cell would respond. Strychnine also abolished the specificity of the local-edge detectors for small spots (that is, a treated cell would respond vigorously to both large and small spots).

Picrotoxin, bicuculline, and strychnine have much less striking effects on the receptive field organization of on- and off-center ganglion cells in the rabbit, especially on the fields of the sustained cells (the X-cells) (Caldwell and Daw, 1978b). For the most part these agents simply increased the spontaneous activity of the cell; they probably were relieving the tonic GABA and glycine inhibition noted above (see also Belgum et al., 1982). In the cat, GABA and glycine antagonists enhanced both spontaneous and light-driven activity of many cells but did not fundamentally alter center–surround organization of the receptive fields (Bolz et al., 1985a,b). In both the cat and the rabbit, picrotoxin appeared to have more effects on transient on- and off-center ganglion cells (the Y-cells) (Caldwell and Daw, 1978b; Kirby and Enroth-Cugell, 1976); and in both species the center–surround balance was sometimes altered in favor of the center response (see also Bolz et al., 1985a,b).

GABAergic amacrine cells also make synapses onto bipolar cells and other amacrine cells in the inner plexiform layer (Vaughn et al., 1981). The effects of GABA and glycine on bipolar and amacrine cell responses have been examined in the mudpuppy retina, which like the rabbit retina has horizontal cells that do not appear to use either GABA or glycine as their transmitter (Miller et al., 1981a). Thus, the observed GABA and glycine effects most likely represent amacrine cell effects. On-center bipolar cell responses in the mudpuppy are somewhat enhanced by GABA antagonists, whereas off-center bipolar cell responses are similarly enhanced by strychnine (Miller et al., 1981a). These results suggest that GABAergic amacrine cells make inhibitory feedback synapses on on-center bipolar cell terminals and glycinergic amacrine cells make inhibitory feedback junctions on off-center bipolar cells. The center–surround antagonism of the two types of bipolar cells in the mudpuppy is not affected by any of these antagonists, which is to be expected if neither GABA nor glycine are horizontal cell neurotransmitters in this animal.

When applied to transient amacrine cells of the mudpuppy, both GABA and glycine reduce the light-evoked responses and decrease substantially the membrane resistance of the cells. GABA and glycine antagonists, on the other hand, enhance the responses of the transient amacrine cells. None of these agents, however, changes the membrane potential of the cells significantly. It appears that GABA and glycine open Cl^- channels in transient amacrine cells, but since chloride is passively distributed across the amacrine cell membrane, no potential change occurs when Cl^- conductance is altered (Miller and Dacheux, 1983). Thus, GABA and glycine appear to inhibit amacrine cells by a "shunting" mechanism; that is, the resistance of the cell is substantially decreased, a change that decreases the effectiveness of excitatory input in depolarizing the cell (see Appendix).

ACETYLCHOLINE   Acetylcholine powerfully excites many types of ganglion cells (Ames and Pollen, 1969; Negishi et al., 1978), whereas other ganglion cell types do not respond to acetylcholine. In most retinas the on–off ganglion cells are affected more by acetylcholine than other types are (Masland and Ames, 1976; Glickman et al., 1982). In the carp, cholinergic antagonists usually blocked all activity of the on–off ganglion cells. This result suggests that the sole excitatory input to these cells is via the acetylcholine-containing amacrine cells (Glickman et al., 1982). In the rabbit, however, similar experiments indicate that there is additional excitatory input to these cells (Masland and Ames, 1976; Ariel and Daw, 1982b). In both the carp and the rabbit acetylcholine excites the ganglion cells following blockade of chemical synaptic input to the cells with agents such as Co^{2+}. This result indicates that acetylcholine acts directly on the ganglion cell membrane (Masland and Ames, 1976; Glickman et al., 1982).

The effects of acetylcholine on ganglion cells in the rabbit retina have been studied by Ariel and Daw (1982a,b), primarily with the

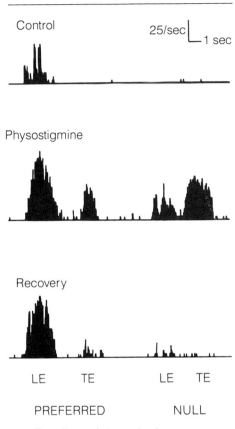

Control

25/sec

1 sec

Physostigmine

Recovery

LE	TE		LE	TE
PREFERRED			NULL	

5.8 The effects of physostigmine on an on-center, direction-sensitive ganglion cell in the rabbit retina. The responses were elicited by a long bar of light moving in the preferred direction (*left responses*) or null direction (*right responses*). During the infusion of physostigmine, an acetylcholinesterase inhibitor (*middle traces*), the response to movement in the preferred direction was enhanced, and the cell also responded strongly to movement in the null direction. LE, response of leading edge of bar; TE, response to trailing edge of bar. From Ariel and Daw (1982b), with permission of the Physiological Society.

use of physostigmine, which is an inhibitor of acetylcholinesterase (the enzyme that breaks down acetylcholine). They found two kinds of effects. First, on X and Y ganglion cells, physostigmine greatly increased spontaneous activity. This tonic acetylcholine input appeared not to be affected by light; rather it seemed simply to contribute to spontaneous activity levels. Second, physostigmine dramatically altered the light responsiveness of the on–off direction-sensitive cells and other motion-sensitive cells. For example, it abolished the direction selectivity of the direction-sensitive cells apparently by overwhelming the GABA inhibition induced by a stimulus moving in the null direction. This effect is shown in Figure 5.8 for an on-center, direction-sensitive unit. Following infusion of physostigmine, the response of the cell to movement in the preferred direction was strongly enhanced, and the enhanced excitatory input permitted responses to occur during movement in the null direction. Curiously, following physostigmine infusion, the on-center cell also began to respond at the offset of illumination; that is, it became on–off. Other types of ganglion cells, particularly those that respond to moving stimuli, also showed on–off responses following physostigmine infusion. The general conclusion from these experiments is that the normal, transient on and off input from cholinergic amacrine cells plays an important role in the excitation of movement- and direction-sensitive ganglion cells.

The acetylcholine-containing amacrine cells have been identified in the rabbit retina (illustrated in Plate 2); they are the so-called starburst amacrines (Famiglietti, 1983; Vaney, 1984; Tauchi and Masland, 1984). These are monostratified cells of two mirror-image types. One, whose perikaryon is in the inner nuclear layer, extends processes into the upper half of the inner plexiform layer; the other, a displaced amacrine cell, has its perikaryon among the ganglion cells and extends processes into the lower half of the inner plexiform layer (Masland and Mills, 1979). The synaptic organization of these amacrines has been studied; they receive both bipolar and amacrine cell input, and they appear to make synaptic contacts exclusively with ganglion cell dendrites (Famiglietti, 1983). Anatomical data thus confirm the physiological evidence that amacrine cells releasing acetylcholine impinge directly onto ganglion cells.

The release of acetylcholine from the rabbit retina occurs in two ways (Masland et al., 1984b). A small tonic release is independent of Ca^{2+}, light, and K^+ levels. It may correspond to the light-independent effects of acetylcholine observed by Ariel and Daw.

The other release is much larger, is Ca^{2+} dependent, and occurs at both onset and offset of light stimulation (Masland and Livingstone, 1976). In other words, the light-evoked release of acetylcholine from the retina is transient and is an on–off release. Evidence indicates that the on-release occurs via the starburst amacrines that sit among the ganglion cells (the displaced starburst amacrine cells) and that the off-release is mediated by the conventionally placed starburst amacrine cells (Masland et al., 1984b).

Finally, as noted in Chapter 4, GABA powerfully inhibits the light-evoked release of acetylcholine from the rabbit retina (Massey and Neal, 1979; Massey and Redburn, 1982), and the effects of GABA can be blocked by both picrotoxin and bicuculline. This finding indicates that GABA is probably acting on specific receptors on the starburst amacrine cell membrane. Furthermore, these GABA antagonists increase the resting release of acetylcholine, a result providing further evidence of an ongoing release of GABA in the retina. It seems likely, therefore, that much of the null inhibition observed in direction-sensitive ganglion cells can be accounted for by serial synaptic interactions between GABAergic and cholinergic amacrine cell processes (see Chapter 4).

MONOAMINES   Both catecholamine-containing and indoleamine-containing amacrine cells occur in most retinas (Plate 5). Dopamine is the principal if not the only, catecholamine present in virtually all retinas (Ehinger, 1976), and dopamine-containing amacrine cells are readily visualized with the Falk-Hillarp fluorescence method (Häggendal and Malmfors, 1963; Ehinger, 1966; Laties and Jacobowitz, 1966). Furthermore, the synthesis of dopamine in the retina has been observed (Da Prada, 1977), release of dopamine by light has been demonstrated (Kramer, 1971), and specific receptors for dopamine in the retina have been identified (Watling and Iversen, 1981). Thus, dopamine fulfills many of the criteria used to establish a substance as a neurotransmitter or neuromodulator in the brain.

In cold-blooded vertebrates and birds the indoleamine, serotonin, appears to be present in the retina, but in mammals physiological quantities of serotonin usually are not found (Ehinger, 1982). Nevertheless, in some mammals (rabbit, pig, and cow) indoleamine-accumulating amacrine cells have been observed. This observation indicates that the indoleamine used by these retinas is something other than serotonin. In other mammals (rat, baboon, and human), no indoleamine-accumulating amacrine cells have been visualized, so indoleamine-accumulating amacrine cells may not be present in these species.

The synaptic organization of the dopaminergic and indoleamine-accumulating amacrine cells has been studied in several species (Dowling and Ehinger, 1978a,b; Ehinger and Holmgren, 1979; Holmgren-Taylor, 1982). Dopaminergic amacrine cells in rabbit, cat, and cynomolgus monkey retinas make synaptic contacts exclusively with other amacrine cells (Figure 5.9). The indoleamine-accumulating neurons in the rabbit and the cat primarily contact bipolar terminals, making many reciprocal junctions with these terminals (Figure 5.9). They also make a few synapses with amacrine cell processes. In the carp, on the other hand, the indoleamine-accumulating amacrine cells make numerous synapses with both amacrine cell processes and bipolar cell terminals. Only in the cat have the dopaminergic or indoleamine-accumulating neurons been

5.9 Summary diagram illustrating the synaptic organization of dopaminergic and indoleamine-accumulating amacrine cells in the rabbit retina. The dopaminergic amacrines (DA) make pre- and postsynaptic contacts only with other amacrine cells and their processes. The amacrine cells that accumulate indoleamines (IA) are pre- and postsynaptic mainly with bipolar terminals (B), although they also make a few synapses with amacrine cell processes. A, amacrine cell; G, ganglion cell. Modified from Dowling and Dubin (1984), reprinted with permission of the American Physiological Society.

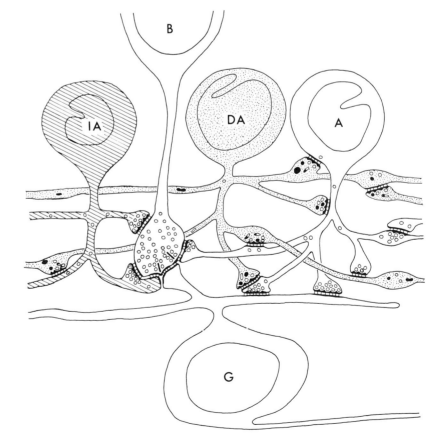

observed to make any direct synaptic contacts with the ganglion cells; and even in this case only a few junctions between indole-amine-accumulating processes and ganglion cells were observed. Thus, the dopaminergic and indoleaminergic amacrine cells appear to operate mainly by modifying the inputs to the ganglion cells and not by directly affecting the ganglion cells.

Little is known concerning the effects of serotonin or other indoleamines on the receptive field properties of ganglion cells. In the rabbit serotonin antagonists applied to the retina reduce on-responses of both on- and off-center cells and enhance off-responses of the off-center cells (Brunken and Daw, 1986). Dopamine effects on the ganglion cell receptive field organization has also been examined in the rabbit retina by the infusion of dopamine antagonists (Jensen and Daw, 1984). These experiments suggested that dopamine modulates the strength of center–surround antagonism, particularly in X-cells, because, following the infusion of the antagonists, surround inhibition was significantly reduced. The dopamine antagonists also reduced the sustained excitation of the on-center X-cells and the leading edge response of on–off, direction-sensitive cells, effects that are not readily understood. A detailed description of the effects of dopamine in the fish retina will be provided in a later section.

PEPTIDES   The evidence that amacrine cells contain peptides, and presumably use them as neurotransmitters or neuromodulators, is mainly immunohistochemical (Brecha et al., 1979; Yamada et al., 1980; Brecha et al., 1984; Bruun et al., 1986). Eight neuropeptides—substance P, enkephalin, somatostatin, neurotensin, glucagon, neuropeptide Y, vasoactive intestinal peptide (VIP), and cholecystokinin (CCK)—have been found in amacrine cells in several species. And radioimmunoassay studies have indicated the presence of other neuropeptides in the retina as well. So other neuropeptides also may be found in amacrine cells.

The peptide-containing amacrine cells have been most extensively studied in the bird retina, and six peptides have been observed immunohistochemically in pigeon and chicken amacrine cells. Figure 5.10 shows a schematic drawing of the peptidergic amacrine cells in the pigeon. None of these types of cells accounts for more than 5 percent of the total number of amacrine cells in the pigeon retina (Karten and Brecha, 1982) and some are very rare; the glucagon- and VIP-containing cells account for 0.8 and 0.5 percent, respectively, of the total amacrine cell population. The

5.10 Summary diagram of pepti-dergic amacrine cells in the pigeon retina. The cells that stain for enke-phalin (ENK), neurotensin (NT), and somatostatin (SS) are very similar in their morphology. Those that contain substance P (SP), vasoactive intes-tinal peptide (VIP), and glucagon (GLU) are different morphologically from the ENK, NT, and SS cells and also from each other. INL, inner nu-clear layer; IPL, inner plexiform layer; GCL, ganglion cell layer. From Stell et al. (1980), reprinted with per-mission from Elsevier Publications Cambridge.

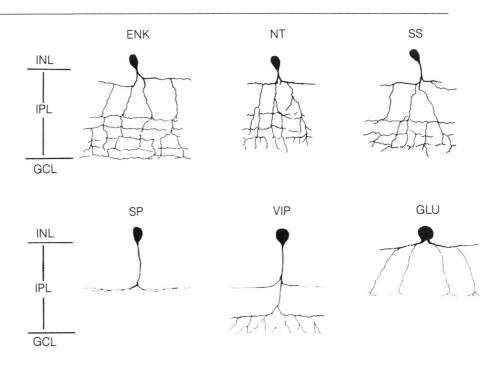

cells containing enkephalin, neurotensin, and somatostatin are re-markably similar morphologically, but are distinct from those con-taining substance P, VIP, or glucagon. The morphological similarity of the enkephalin-, neurotensin-, and somatostatin-containing amacrine cells raises the question of co-localization of peptides in these cells; indeed, this phenomenon has been observed in the chicken retina (Li et al., 1985). About 40 percent of the neuroten-sin-containing amacrine cells also stain for enkephalin; and in a small percentage of cells both neurotensin and somatostatin are found (H.-B. Li, personal communication). It may be that some amacrine cells contain all three peptides, but this has yet to be shown.

The situation is further complicated by evidence showing that about 50 percent of the neurotensin cells accumulate glycine, a finding suggesting that these two substances are co-localized in the same cell (Watt et al., 1985a). Glycine also accumulates in about 10 percent of cells that show enkephalin immunoreactivity and in a small number of the somatostatin-containing cells. And some of the enkephalin-staining cells accumulate GABA (Watt et al., 1984). Thus, it is conceivable that one amacrine cell in the chicken could

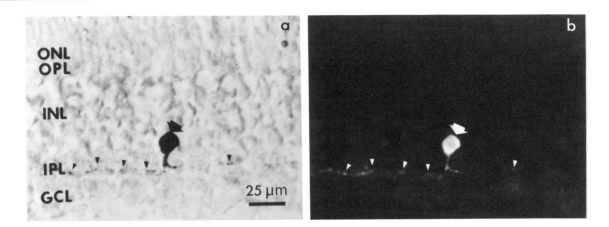

5.11 Photomicrographs showing the co-localization of neurotensin and substance P in the same amacrine cell in the goldfish retina. The same section was exposed both to an antibody that was linked to peroxidase and reacted with neurotensin (*left*) and to an antibody that was linked to a fluorescent marker and reacted with substance P (*right*). The neurotensin (peroxidase) staining is seen by conventional light microscopy (*left*), the substance P (fluorescence) labeling by fluorescence microscopy (*right*). The arrowheads point to varicosities that stain for both peptides. ONL, outer nuclear layer; OPL, outer plexiform layer; INL, inner nuclear layer; IPL, inner plexiform layer; GCL, ganglion cell layer. From Li et al. (1986), reprinted with permission of Elsevier Science Publishers.

contain five neuroactive substances—neurotensin, enkephalin, somatostatin, GABA, and glycine.

Co-localization of neuropeptides with conventional transmitters or other neuropeptides has been observed in other retinas. In turtles, amacrine cells exhibiting neurotensin immunoreactivity show glycine uptake (Weiler and Ball, 1984); and in the goldfish, all the cells that show substance P staining also show neurotensin-like immunoreactivity and vice versa (Figure 5.11). In certain of these cells in the goldfish, the coexistence of the two immunoreactivities can be localized to the same varicosities, an observation suggesting that both peptides are likely to be released at the same synaptic site (Li et al., 1986).

Relatively little is known about the synaptic organization or function of the peptide-containing amacrine cells in the retina. But the amacrine cells containing substance P (and neurotensin) in the goldfish have been carefully examined by electron microscopy. These studies have shown that the processes of these cells receive

5.12 Computer-generated display of the spike responses of an on–off ganglion cell in the carp retina. Each blip represents a single spike, and successive sweeps are positioned one below another. The light stimuli (0.4-mm spot of intensity log $I = -2.75$) were applied every 10 sec. Following the application of Co^{2+} (*top arrow*), the light-evoked responses disappeared and the spontaneous activity of the cell decreased substantially. Following the subsequent application of substance P (SP; 1.5×10^{-6} *M; third arrow from top*), the cell was excited. A second, smaller application of SP (*below*) caused little effect, and some light-evoked change in firing appeared to be reappearing in the last few traces. Modified from Glickman et al. (1982), with permission of Elsevier Science Publishers.

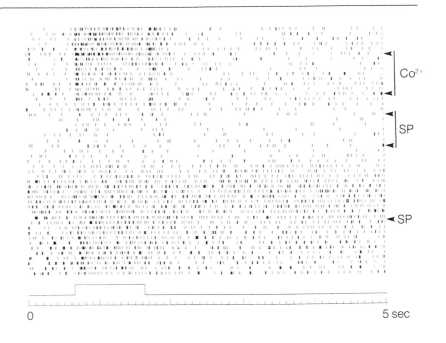

approximately equal input from bipolar and amacrine cell terminals and that they make approximately equal numbers of synapses on amacrine and ganglion cell processes (Yazulla et al., 1985). Other studies have shown that substance P has effects on many ganglion cells in the carp retina (Glickman et al., 1982). About half of the on- and on–off cells were excited by substance P but some (10 percent) were inhibited. On the other hand, substance P tended to inhibit many of the off-center cells (approximately 35 percent), although it excited others (approximately 20 percent). Whether all of these effects on the ganglion cells are direct is not clear. In several cases substance P was applied to a cell after synaptic transmission in the retina was blocked by Co^{2+}, and such experiments showed that direct excitation of ganglion cells by substance P did occur. Because Co^{2+} usually blocked both light-evoked and spontaneous activity of the ganglion cells, it was impossible to determine whether the inhibitory effects of substance P were direct or not. Figure 5.12 shows an example of an excitatory effect of substance P on a carp ganglion cell following the application of Co^{2+} to the retina. Initially the cell had a good light response and moderate spontaneous activity. Following Co^{2+} application the light response was lost and the spontaneous activity of the cell decreased substantially. The subsequent application of substance P induced substantial firing of the cell. This effect was prolonged, often lasting up to 15 min after application of the drug, and the latency between the application and detectable effect of substance P was two to three times longer than that seen with acetylcholine on similar cells.

The enkephalin-containing cells and their synaptic organization in the chicken retina have also been studied in detail. It has been shown, first of all, that met-enkephalin can be synthesized in the chicken retina and that newly synthesized met-enkephalin is released by depolarizing levels of K^+. The addition of Co^{2+} to the incubation medium blocked much of the enkephalin release, a result suggesting that the release mechanism is Ca^{2+} dependent (Su et al., 1985). Electron microscopic studies have shown that these cells appear to receive much of their input from other amacrine cells, although some bipolar input was seen. The output of these cells appeared to be mainly onto other amacrine cells, but many of the postsynaptic processes could not be identified (Watt et al., 1985b). A most interesting observation made in both goldfish and chicken retinas was that enkephalin significantly inhibited the release of GABA from amacrine cells (Djamgoz et al., 1981; Watt et al., 1984). Earlier, we noted that in the rabbit retina, GABA inhibited the release of acetylcholine. An obvious question is whether the inhibition of acetylcholine release by GABA can be modulated by enkephalin.

No studies on the electrophysiological effects of enkephalins on ganglion cells in the chick have appeared. In fishes opiates enhance both light responses and spontaneous activity of on-center cells and inhibit both light responses and spontaneous activity of off-center cells (Djamgoz et al., 1981). In the mudpuppy, on the other hand, exogenously applied enkephalins and other opiates generally inhibited all ganglion cell responses (Dick and Miller, 1981).

Interplexiform Cells

The interplexiform cells were first detected in the vertebrate retina because of their pharmacology. That is, in teleost fishes, the interplexiform cells contain dopamine, which can be visualized by fluorescence microscopy when the tissue is processed by the Falk-Hillarp technique (Plate 6) (Ehinger et al., 1969). Subsequently, these cells were visualized by other methods, particularly Golgi staining, and have been observed in a variety of species (Gallego, 1971; Boycott et al., 1975; Oyster and Takahashi, 1977). But not all interplexiform cells contain dopamine; indeed, present evidence suggests that dopaminergic interplexiform cells may be in the minority. The interplexiform cells in teleost fishes and New World monkeys appear to be mainly dopaminergic, and there appear to be some dopaminergic interplexiform cells in human and cat retinas (Frederick et al., 1982; Oyster et al., 1986). But the majority of the interplexiform cells in the cat appear to use another transmitter, perhaps GABA (Nakamura et al., 1980). The interplexiform cells in the skate also appear to be GABAergic (Brunken et al., 1986), but glycine uptake studies in frogs, toads, and goldfish indicate that there may be glycinergic interplexiform cells in these animals (Rayborn et al., 1981; Kleinschmidt and Yazulla, 1984; Marc and Liu, 1984). The identity of the interplexiform cell transmitter in other retinas is unknown.

The dopaminergic interplexiform cells in teleost fishes have been studied extensively and they will be discussed in some detail in the next section. Much less is known about the presumed GABAergic and glycinergic interplexiform cells. It is of interest in this regard to note that in the skate GABA strongly depolarizes horizontal cells (Lasater et al., 1984) and in *Xenopus* toad glycine strongly depolarizes the horizontal cells (Stone and Witkovsky, 1984), effects perhaps related to interplexiform cell activity in this species.

Case Study of a Neuromodulator: Dopamine in the Teleost Retina

As already noted, two classes of substances appear to be released from chemical synaptic terminals: neurotransmitters and neuromodulators. In the retina, as elsewhere, much more is known about the neurotransmitters, and most of the neuroactive substances in the retina whose effects have been studied in some detail fall into the category of neurotransmitters. These substances include L-glutamate, GABA, glycine, and acetylcholine. It has been shown that these agents all act directly on postsynaptic retinal cells, altering membrane permeability and inducing rapid changes in membrane potential or membrane resistance or both. The action of dopamine in the teleost fish retina, on the other hand, is quite different. It does not directly affect the membrane potential or the membrane conductance of the horizontal cells, its primary target in the teleost retina (Dowling et al., 1983). Rather, it activates an intracellular enzyme system, and the principal physiological effects induced by dopamine in the teleost retina appear to be mediated biochemically. The physiological changes induced by dopamine are multiple and long-lasting, with durations of 15 min or more after a short pulse of dopamine has been applied to the retina or to isolated cells.

Figure 2.9a and Plate 6 show light micrographs of teleost retinas, illustrating the interplexiform cells, which are the cells containing dopamine in these animals. Figures 3.12 and 3.13 show synapses made by the interplexiform cells in the goldfish and a summary diagram of the synaptic organization of the cells in this teleost. As noted earlier, the interplexiform cells appear to be centrifugal neurons in the retina. That is, they receive all of their input in the inner plexiform layer, mainly from amacrine cells, whereas the bulk of their output is made in the outer plexiform layer. Some synapses made by these neurons onto amacrine cell processes in the inner plexiform layer are also observed. In the outer plexiform layer of fishes, the great majority of synapses made by the interplexiform cells are onto the external (cone-related) horizontal cells but some junctions are also observed onto bipolar cells. Never are interplexiform cell synapses seen onto receptor terminals (Dowling and Ehinger, 1978a).

Because horizontal cells in fishes are particularly large and are richly innervated by the dopaminergic interplexiform cell terminals, it has been possible to analyze the action of dopamine on these neurons in some detail. Much less is known about the action of dopamine on the bipolar and amacrine cells. In the first studies on the action of dopamine on teleost horizontal cells (Figure 5.13), pieces of goldfish retina were used and intracellular recordings showed that dopamine usually depolarized the horizontal cells by 5–10 mV and reduced the light-evoked responses of the cells by up to about 40 percent (Hedden and Dowling, 1978). Furthermore, dopamine decreased the antagonistic surround responses of both bipolar cells and cones, effects attributed to the decrease in responsiveness of the horizontal cells induced by dopamine (Figure 5.13). Dopamine also increased the center response of bipolar cells and depolarized transient amacrine cells.

Subsequently, Negishi and Drujan (1979) found that dopamine also significantly altered the receptive field size of the horizontal cells in fishes. Following dopamine application to a retina, the response of a horizontal cell to a spot of light increased significantly, but the response to an annulus of light decreased by about half. Figure 5.14 shows an experiment illustrating this effect, although in this case the two stimuli were a spot of light and full-field illumination. Initially, the spot and the full-field stimulus were adjusted in intensity to give responses of approximately equal amplitude.

5.13 The effects of dopamine on intracellularly recorded responses of a horizontal cell (a), depolarizing bipolar cell (b), and red cone (c).

a Dopamine depolarized the horizontal cell and reduced its light responsiveness. The effects of dopamine were blocked by the antagonist phentolamine.

b Dopamine hyperpolarized the bipolar cell; and following dopamine application the depolarizing response to center illumination (raised abscissa) was increased, while the hyperpolarizing response to annular illumination (decreased abscissa) was decreased.

c Dopamine caused no change in the resting membrane potential of the cone. It did affect the waveform of the response, however, making the response simpler and squarer. The change in waveform reflects a decrease of horizontal feedback onto the receptor. Log I, relative flash intensity. Modified from Hedden and Dowling (1978), with permission of the Royal Society.

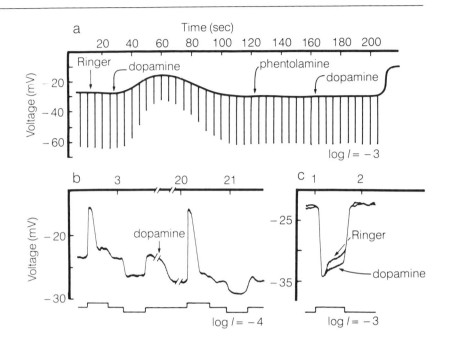

(Because horizontal cells are much more responsive to large fields than to spots of light, the intensity of the spot was considerably brighter than was the full-field stimulus.) Following superfusion of the retina with dopamine for approximately 30 sec at a concentration of 25 μM, the cell depolarized slightly (~2.5 mV), the response to the spot increased by over threefold, and the response to full-field illumination decreased by about 35 percent. These effects were long-lasting: 7–8 min after dopamine infusion ceased, the change in receptive field size was still obvious. Negishi and Drujan suggested on the basis of such experiments that lateral propagation of signals between horizontal cells was depressed by dopamine. Because it is well established that horizontal cells are electrically coupled, these experiments immediately suggested that dopamine was somehow also altering the coupling between horizontal cells.

Figure 5.15 shows in graphic form the effects of dopamine on the receptive field profile and responsiveness of horizontal cells in the isolated carp retina. Response amplitude in millivolts is plotted versus spot size, and these data show that with small spots (0.2–0.8 mm diameter) dopamine increases response amplitudes, but with large spots (3.2–8 mm) and full-field illumination dopamine significantly decreases response amplitudes. The increase in re-

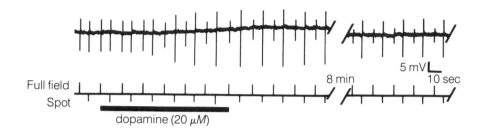

5.14 Intracellular records of carp horizontal cell responses to spot and full-field white-light stimuli before, during, and after the addition of dopamine (20 μ*M*) to the superfusion medium. The spot (0.8 mm in diameter) and full-field stimuli were presented as an alternating pair and adjusted before dopamine application to generate responses of approximately equal amplitudes. Dopamine caused the responses to the spot stimuli to increase in amplitude and the responses to the full-field stimuli to decrease in amplitude. Recovery from these drug effects required an average of about 15 min. Note that dopamine also caused the horizontal cell to depolarize slightly. Modified from Mangel and Dowling (1985), with permission; copyright 1985 by the AAAS.

sponses to small spot stimuli can be explained by decreased electrical coupling between horizontal cells. But the decreased responsiveness to large spot and full-field stimuli cannot be explained on this basis. With full-field stimuli, little current flow should occur between adjacent horizontal cells, because they all are receiving equal input from the receptors and should be at the same potential. Thus, altering the strength of coupling between cells should not have effects on response amplitudes elicited with full-field stimuli. Figure 5.15 indicates, therefore, that dopamine induces two independent effects on horizontal cells: decreased responsiveness of the cells to large light stimuli and decreased electrical coupling between adjacent horizontal cells.

None of the preceding experiments provided any information concerning the mechanisms underlying the effects of dopamine on the horizontal cells, but it was assumed that dopamine had direct effects on the horizontal cell membrane (Hedden and Dowling, 1978). When dopamine was applied to isolated fish horizontal cells, however, it was found that at physiological concentrations (< 300 μ*M*) dopamine altered neither the membrane potential nor the membrane resistance of these cells (Lasater and Dowling, 1982). (Why dopamine usually depolarizes horizontal cells in the intact retina but does not affect membrane potential or resistance in isolated cells is discussed later.) How, then, does dopamine exert its effect on horizontal cells?

In many parts of the brain, dopamine is known to interact with specific membrane receptors that are linked to adenylate cyclase, the enzyme that converts ATP to cyclic AMP. Cyclic AMP serves as an intracellular messenger in many cells and is capable of affecting many aspects of cell function. It has been known for some time that dopamine activates adenylate cyclase in the retina (Brown and Makman, 1972), and a highly specific dopamine-sensitive adenyl-

5.15 Average carp horizontal cell response amplitudes as a function of stimulus spot diameter. The stimuli were centered on receptors in the middle of the cell's receptive field and were at an intensity that generated a half-maximal response when a full-field stimulus was used. Dopamine application caused average response amplitudes to small spot stimuli to be significantly larger and average response amplitudes to large spot and full-field stimuli to be significantly smaller. Modified from Mangel and Dowling (1985), with permission; copyright 1985 by the AAAS.

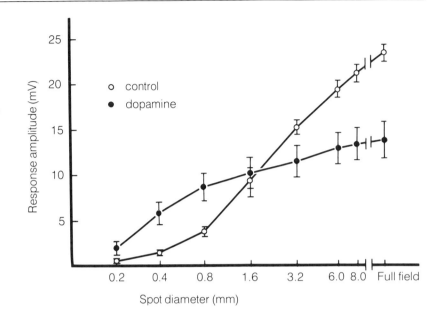

ate cyclase is present in the fish retina (Watling and Dowling, 1981; Dowling and Watling, 1981). Evidence that the dopamine receptors on horizontal cells are linked to adenylate cyclase was provided by preparing fractions of isolated horizontal cells (and other cell types) and determining cyclic AMP content after exposure to dopamine (Van Buskirk and Dowling, 1981). Figure 5.16 shows the results of such experiments. Fractions containing rod photoreceptors demonstrated substantial basal levels of cyclic AMP, but their cyclic nucleotide content was unaffected by dopamine in the incubating medium. Conversely, fractions enriched in horizontal cells showed very low basal levels of cyclic AMP, but then showed a graded increase in cyclic AMP levels with increasing concentrations of dopamine in the incubating medium. About 10 μM of dopamine were required to half-saturate the system, the same concentrations of dopamine required to half-saturate cyclic AMP accumulation in the intact retina. These experiments showed not only that there are dopamine receptors linked to adenylate cyclase on isolated horizontal cells but also that these receptors are not altered significantly by the isolation of the horizontal cells from the retina. Thus, the failure of dopamine to alter the membrane potential or membrane resistance of isolated horizontal cells was not due to a loss of dopamine receptors from the cells.

5.16   Effects of various concentrations of dopamine on cyclic AMP content of purified fractions of isolated horizontal cells and rod cells from the carp retina. All fractions were preincubated in Ringer solution containing 2 mM isobutylmethylxanthine (a phosphodiesterase inhibitor) for 3 min and then incubated in Ringer solution containing dopamine for 5 min. Results expressed as femtomoles of cyclic AMP per horizontal cell are based on the assumption of 3,000 horizontal cells per fraction. Standard errors of the mean are provided for those points based on three or more experiments. Dopamine caused a graded increase in cyclic AMP content in the horizontal cell fractions, but not in the rod cell fractions. From Van Buskirk and Dowling (1981), with permission.

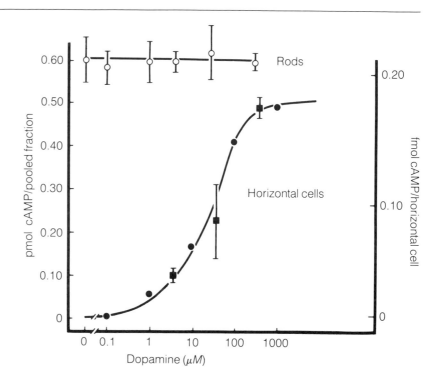

Substantial evidence that dopamine alters electrical coupling between horizontal cells and that dopamine exerts this effect via cyclic AMP has come from a variety of experiments. For example, it has been shown in both fish and turtle retinas that the application of dopamine, dibutyryl-cyclic AMP (a stable cyclic AMP analogue), and compounds that increase intracellular levels of cyclic AMP all narrow the receptive fields of horizontal cells and decrease the diffusion of dye between the cells (see Plate 3) (Teranishi et al., 1983, 1984; Piccolino et al., 1984). Destruction of the interplexiform cells with the neurotoxin 6-hydroxydopamine, on the other hand, broadens the receptive field profile of horizontal cells and enhances dye diffusion between horizontal cells (Teranishi et al., 1983; Cohen and Dowling, 1983). Direct evidence that dopamine and cyclic AMP decrease the conductance of the electrical junctions between horizontal cells has come from studies on pairs of electrically coupled horizontal cells in culture (Lasater and Dowling, 1985a,b). Following the application of a brief pulse of dopamine onto two such cells, or the injection of cyclic AMP into one cell, the resistance of the gap junctional membrane was observed to increase, often by

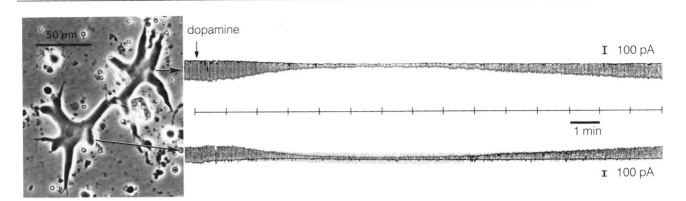

5.17 Effects of dopamine on an electrically coupled pair of white perch horizontal cells maintained in culture (*left*). Both cells were voltage-clamped at −60mV with patch electrodes and current pulses were applied to the driver cell (*lower trace*) to shift the membrane potential +20 mV. Ringer solution containing dopamine was applied briefly (0.5 sec) to the cell pair. The cells uncoupled, as shown by the decrease in magnitude of the current pulses required to depolarize the driver cell by 20 mV, which reflects the increase of resistance of the driver cell; and the decrease in the magnitude of the current pulses passed into the follower cell (*upper trace*), which reflects the decreased conductance of the junctional membrane. Modified from Lasater and Dowling (1985a), with permission.

a factor of ten or more. Figure 5.17 shows such an experiment on a pair of cultured perch horizontal cells. The membrane potential of both cells was first maintained (voltage-clamped) at −60 mV, and then the potential of one of the cells, the driver cell (lower trace), was shifted by +20 mV with current pulses. In response to dopamine (arrow), the cells uncoupled, as shown by the decrease in magnitude of the pulses recorded in both cells. The pulses in the follower cell (upper traces) became smaller because of the decreased conductance of the gap junctional membrane (less current was flowing into the follower cell); the magnitude of the pulses in the driver cell (lower trace) decreased because of the increased junctional resistance between the cells (the resistance of the driver cell was increased, less current was needed to depolarize the cell by 20 mV). Before the application of the dopamine pulse, a junctional resistance of ~80 megohms was measured. After drug application there was a latency of about 30 sec before uncoupling began. The cells were maximally uncoupled after about 4 min, at which point the junctional resistance had risen to 660 megohms. Coupling between the cells remained low for about 1 min and then slowly in-

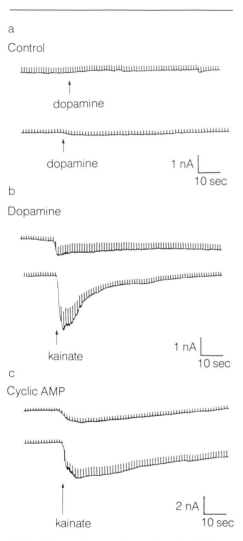

a
Control

dopamine

dopamine   1 nA
                      10 sec
b
Dopamine

kainate   1 nA
                     10 sec
c
Cyclic AMP

kainate   2 nA
                     10 sec

5.18 Effects of dopamine and cyclic AMP on ionic conductances mediated by kainate (a glutamate agonist) in perch horizontal cells maintained in culture.

a Short (~0.5 sec) pulses of dopamine (200 μM) to horizontal cells caused no significant changes in membrane conductance.

creased. After about 14 min, the junctional resistance had recovered to 100 megohms. For other cell pairs tested in this way, or following the injection of cyclic AMP into one of the cells, junctional resistance increased from control levels of 20–60 megohms up to 300–700 megohms.

How cyclic AMP modifies gap junctional conductance between horizontal cells is not known. In many cyclic AMP-mediated systems, protein phosphorylation is involved (Greengard,1978). The cyclic AMP activates specific kinases that in turn phosphorylate particular proteins. Phosphorylation of proteins is known to alter the proteins' properties, and it may be that phosphorylation of gap junctional membrane proteins alters the conductance of the junction. Recent experiments have shown that the injection of the catalytic subunit of cyclic AMP-dependent kinase into horizontal cells uncouples pairs of horizontal cells maintained in culture, a finding supporting this hypothesis (E. M. Lasater, personal communication).

As noted earlier, dopamine also decreases the responsiveness of horizontal cells in the intact retina to large spot or full-field stimuli. Experiments on single perch horizontal cells in culture suggest how dopamine, acting via cyclic AMP, could alter the responsiveness of horizontal cells to light (Knapp and Dowling, 1987). Both dopamine and cyclic AMP significantly enhance ionic conductance changes elicited in isolated horizontal cells by L-glutamate and kainate, agents that mimic the photoreceptor transmitter (Figure 5.18). The application of dopamine itself (200 μM) to isolated horizontal cells induces no significant change in membrane conductance (top traces in Figure 5.18). The lower traces show the conductance changes induced in these cells by kainate (50 μM) before and a few minutes after the application of a pulse of Ringer solution containing 200 μM dopamine or 500 μM cyclic AMP to the cells. In both cases, the currents induced by kainate were enhanced by about 2.5-fold. This effect of dopamine was blocked by haloperidol, a dopamine antagonist. In rod horizontal cells, which do not receive synaptic input from dopaminergic interplexiform cells, neither dopamine nor cyclic AMP affected kainate-induced conductances.

An explanation for the decrease in light responsiveness of horizontal cells following dopamine application to the intact retina goes as follows: light reduces the flow of transmitter from the receptors, thereby hyperpolarizing horizontal cells. Dopamine, by enhancing the potency of the photoreceptor transmitter, renders the

5.18 *(cont.)*

b and c Kainate- (50 μ*M*) induced conductances before (*upper traces*) and a few minutes
after (*lower traces*) the application of a pulse
of dopamine (**b**) or cyclic AMP (**c**) to horizontal
cells. Both dopamine (200 μ*M*) and cyclic
AMP (500 μ*M*) enhanced kainate-mediated
conductance by two- to three-fold. From
Knapp and Dowling (1987), reprinted with permission from *Nature* 325:437–439, copyright ©
1987, Macmillan Journals Ltd.

light-induced reductions of transmitter release less effective, and so decreases the amplitude of the cell's response to light. These findings may also provide an explanation for the paradoxical observations that dopamine usually depolarizes horizontal cells in the intact retina but has no consistent effect on the membrane potential or resistance of isolated horizontal cells. As discussed earlier, the ongoing release of photoreceptor transmitter continually depolarizes horizontal cells. Dopamine, by facilitating the action of the transmitter, would depolarize horizontal cells further in the intact retina, when transmitter is present, but would have no effects on isolated cells maintained in culture, when transmitter is absent.

Other effects have also been reported to occur in horizontal cells following the application of dopamine or cyclic AMP-promoting agents to the retina. For example, dopamine and cyclic AMP analogues substantially decrease both the dark-release and glutamate-induced release of GABA from horizontal cells (Yazulla and Kleinschmidt, 1982; Yazulla, 1985). Thus, it would appear that dopamine and cyclic AMP induce a variety of changes within the horizontal cells.

Overall, the action of dopamine on the horizontal cell is to decrease the effectiveness of the cell. As described in Chapter 4, the horizontal cells mediate lateral inhibitory effects in the outer plexiform layer of the retina and form the antagonistic surrounds of the bipolar and receptor cells. Decreasing light responsiveness of the cell, shrinking its receptive field size, and depressing its release of neurotransmitter are all effective ways to lessen horizontal cell influence. Thus, a decrease in bipolar and receptor surround responses are to be expected following dopamine application and such changes have been observed (Figure 5.13). Following dopamine application to the catfish retina, the receptive field size of the horizontal cells decreases to a size smaller than the center field of the bipolar cell, and, consequently, no antagonistic surround is observed in the bipolar cell response (K.-I. Naka and J. E. Dowling, unpublished observations).

What might be the significance of the modulation of lateral inhibition and surround antagonism by dopamine and the interplexiform cells in the retina? It has long been known that following prolonged periods of time in the dark the antagonistic surrounds of many ganglion cells are reduced in strength or even eliminated. This was first shown by Barlow et al. (1957) in the cat, and a similar phenomenon has been reported to occur in the frog (Donner and Reuter, 1965) and the rabbit (Masland and Ames, 1976). This change in receptive field organization does not relate to a switch from cone to rod vision, and what mechanism underlies this alteration in center–surround organization is unknown. An obvious speculation is that the interplexiform cells or dopamine or both play such a role and regulate the strength of lateral inhibition and center–surround antagonism in the retina as a function of adaptive state (Hedden and Dowling, 1978).

Evidence in favor of this mechanism has been provided by examining the receptive field profiles of horizontal cells in the carp retina following short (30–40 min) and long (100–120 min) periods of time in complete darkness (Mangel and Dowling, 1985). The receptive field profile of horizontal cells in retinas maintained for long periods of time in the dark is distinctly different from those profiles measured on horizontal cells in retinas maintained in the dark for short periods of time (Figure 5.19). The profiles from retinas in prolonged darkness resemble very much those of short-term dark-adapted horizontal cells exposed to dopamine. Indeed, the application of exogenous dopamine to retinas maintained for long periods in the dark causes no effects on horizontal cell receptive fields, a finding suggesting that after a prolonged period of time in the dark dopamine is released tonically from the retina, that is from the interplexiform cells.

Little is known about the mechanisms regulating dopamine release from the interplexiform cells in the fish. No transmitter candidate that releases dopamine from the fish retina has been found. On the other hand, GABA antagonists readily release dopamine from the retina, a finding suggesting that the interplexiform cells are under a strong tonic inhibitory control by GABAergic amacrines (O'Connor et al., 1986). It is possible that the release of dopamine is entirely regulated in this way and that there is no excitatory input to these cells; but this remains to be shown. It is interesting to note that the application of GABA antagonists to the intact retina narrows the receptive fields of horizontal cells in both turtles and fishes, not by an action of these agents on the horizontal cells, but because they release dopamine from the interplexiform cells (Piccolino et al., 1982; Negishi et al., 1983).

In summary, dopamine and the interplexiform cells induce multiple effects on fish horizontal cells, most of which serve to depress the activity of these neurons. These effects appear to be mediated biochemically, within the cell, and are long-lasting. Thus, the horizontal cell, which shapes the receptive fields of the retinal neurons through inhibitory interactions with bipolar and receptor cells is

5.19 a Average carp horizontal cell response amplitudes as a function of stimulus spot diameter following 30–40 min in the dark (control) and following 100–120 min in the dark (prolonged darkness). Prolonged darkness, like dopamine, caused averaged response amplitudes to small spot stimuli to be significantly larger and response amplitudes to large spot stimuli to be significantly smaller. Modified from Mangel and Dowling (1985).

b Intracellular carp horizontal cell responses to spot and full-field white-light stimuli following dopamine application to a retina kept in the dark for a prolonged period (100–120 min). The experimental protocol was identical to that used in Figure 5.14; but, under these conditions of prolonged dark adaptation, dopamine caused no significant effects on the relative response size to spot and full-field stimuli. Modified from Mangel and Dowling (1985), with permission; copyright 1985 by the AAAS.

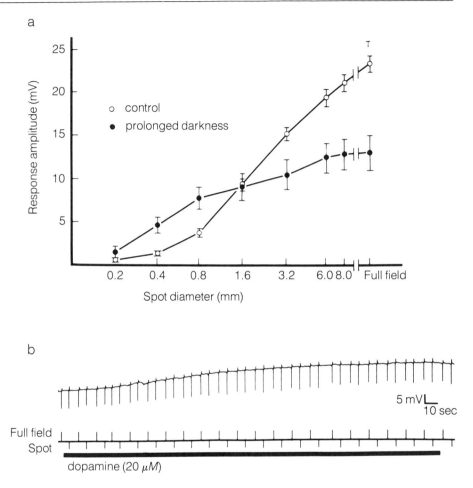

itself under the influence of another class of neuron, the interplexiform cell, which modulates the inhibitory interactions mediated by the horizontal cell. There are, therefore, two levels of control exerted on the receptive fields of the receptor and bipolar cells, and ultimately on all of the retinal neurons; a direct one mediated by the horizontal cells and an indirect one, mediated by the interplexiform cells.

The Electroretinogram and Glial Responses

6.1 a Schematic drawing showing the conditions under which an electroretinogram may be evoked and recorded. An active electrode is placed upon the cornea of the eye and an indifferent electrode on the forehead. The response, elicited by a flash of light projected onto the retina, is amplified and observed on an oscilloscope. Redrawn from Dowling and Ripps (1976b), with permission of Woods Hole Oceanographic Institute.

WHEN a low-resistance electrode, such as a Ringer-soaked cotton wick or a length of silver–silver chloride wire, is positioned on the corneal surface of an eye and a second (reference) electrode is placed elsewhere on the head, voltage changes are recorded each time the eye is stimulated with a flash of light (Figure 6.1). Such potentials are called field potentials or gross potentials. They reflect the summed electrical activity of populations of cells and are usually recorded from the surface of a tissue (see Armington, 1974). The field potential recorded from the eye is called an electroretinogram (ERG). Field potentials can also be recorded from the surface of the skull—the electroencephalogram (EEG)—and from the skin overlying the heart—the electrocardiogram (EKG).

The electroretinogram was among the earliest recorded biological potentials. Credit for its discovery is shared by Frithiof Holmgren of Sweden and James Dewar and John McKendrick of Scotland. Holmgren was the first to record an ERG (1865), but he misinterpreted its significance. He believed that the electrical deflections he observed at the onset and cessation of illumination when recording from a fish eye represented the summed discharge of action potentials arising from the optic nerve. Only later did he come to realize the electrical responses he recorded arose from within the retina itself. Dewar and McKendrick (1873), on the other hand, independently suggested that the potentials they observed arose from within the retina.

The ERG is easy to record from intact eyes of both animals and humans in awake or anesthetized subjects. ERGs can also be readily recorded from excised eyes, eyecup preparations, or isolated retinas with electrodes placed across the eye, eyecup, or retina, respectively. The potentials evoked in different species with reasonably bright and large-field stimuli range from a few hundred microvolts to 1–2 millivolts. Because recordings tend to be relatively stable for long periods of time, ERG potentials can be studied quantitatively under a wide variety of conditions.

The ERG has proved to be a very useful tool for evaluating retinal and visual function. It has a secure place in clinical diagnosis

6.1 *(cont.)*

b Electroretinograms (ERGs) recorded from the all-rod retina of the skate (*top*) and from the cone-dominated retina of the ground squirrel (*bottom*). The ERG from the skate was recorded from an isolated eyecup preparation; the ERG from the ground squirrel was recorded from the intact eye of an anesthetized animal. The skate ERG was elicited with a 0.2-sec flash; the ground squirrel ERG was elicited with a 1.0-sec flash. Both responses show a- and b-waves. However, the rod ERG shows a prominent slow c-wave while the cone ERG shows a d-wave at the offset of illumination. *Top* from Dowling and Ripps (1972), reprinted with permission of the Rockefeller University Press; *bottom* redrawn from Green and Dowling (1975), with permission of Alan R. Liss.

and is useful for a number of kinds of studies in visual physiology, providing information concerning the kinds and spectral sensitivities of the photoreceptor elements in a retina and serving as an index of the absolute sensitivity of the retina to photic stimuli. The analysis of light and dark adaptation using the ERG to determine changes in light sensitivity has been particularly profitable (Chapter 7).

When elicited with a bright stimulus, the ERG is a complex potential that consists of several major components and several minor ones. Figure 6.1b shows electroretinograms recorded from the all-rod retina of the skate and from the virtually all-cone retina of the ground squirrel (Dowling and Ripps, 1970; Green and Dowling, 1975). In both records three major waves are observed, although they are not the same in each case. Both responses show an initial potential, the a-wave, whose polarity is negative relative to the cornea. This is followed in both records by a corneal-positive potential, the b-wave. In the rod-mediated ERG a second, much slower corneal-positive potential is seen: the c-wave. The c-wave is not usually observed in the cone ERG (but see Matsuura et al., 1978). There is in the cone ERG, however, a prominent off-potential at the cessation of the stimulus (the d-wave), which is not seen in rod ERGs. Rod ERGs show only a small negative dip in the record at the cessation of illumination (see Figure 6.2). In retinas that contain both rods and cones all four major potentials may be seen under the right conditions (see Figure 6.15).

There are other ERG potentials that can be detected under various conditions. With relatively bright, repetitive flashes, oscillatory wavelets are superimposed on the b-wave, and these potentials appear to be separate from the b-wave. With very high intensity flashes, very short latency potentials can be distinguished and are separable from the initial, negative a-wave evoked with light flashes of ordinary intensity. And, finally, it is possible, particularly with pharmacological manipulations, to observe a very slow corneal-negative potential that far outlasts the b-wave.

ERG Components and Their Cellular Origins

Understanding the components of the ERG, the cellular origins of the various ERG waves, and how the potentials are generated is of obvious importance, and substantial progress has been made in this direction. One of the first component analyses was carried out by Granit (1933), who worked with dark-adapted, rod-dominated cat

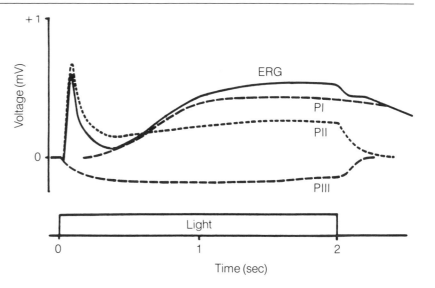

6.2 Component analysis of the ERG recorded from the cat. The continuous line represents the response of the eye in the dark-adapted state; the dotted lines represent the three basic components, PI, PII, and PIII, that underlie much of the response. From Granit (1933), reprinted with permission of the Physiological Society.

eyes. He observed the changes that occurred in the ERG during ether narcosis and postulated the existence of three separate components of the ERG. The PI component, which appeared to give rise to the c-wave, was most sensitive to ether narcosis and disappeared first, leaving the a- and b-waves apparently intact. The PII component, which gives rise to the b-wave, was the next to disappear, leaving the negative PIII component, which was most resistant to the ether. Figure 6.2 shows Granit's analysis of the cat ERG.

Virtually all subsequent analyses of the ERG have started from Granit's scheme of three basic ERG components. It is now clear, however, that his PIII component consists of two separate components that arise from different retinal cells and that, as already noted, there are additional minor potentials not observed or analyzed by Granit. Of particular interest, and of some surprise, has been the realization that most of the potentials giving rise to Granit's three components are generated not directly by the retinal neurons but by nonneural elements, namely, the Müller (glial) cells and the pigment epithelium. The early, fast subcomponent of PIII is the only exception; it derives from the photoreceptors. However, the potentials generated in the Müller and pigment epithelial cells appear to arise from light-evoked changes in K^+ levels that occur as a result of neuronal or photoreceptor cell activity, so the ERG components do ultimately reflect neural cell activity. Studies comparing relative changes in light sensitivity of individual ganglion cells and

ERG

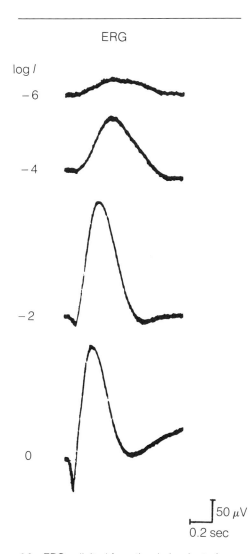

log *I*

−6

−4

−2

0

50 μV

0.2 sec

6.3   ERGs elicited from the dark-adapted skate eyecup over a range of light intensities (log *I*). At low intensities (log *I* = −6 and −4), only the b-wave is evident; at higher intensities (log *I* = −2 and 0), an a-wave precedes the b-wave. At the highest light intensity (log *I* = 0), the beginning of a c-wave can be seen following the b-wave. From Dowling and Ripps (1970), reprinted with permission of the Rockefeller University Press.

of the b-wave in the skate during both light and dark adaptation show virtually identical results (Dowling and Ripps, 1970). Thus, the b-wave, although not arising directly from the retinal neurons, shows adaptive properties very similar to those recorded from the proximal retinal neurons (Green et al., 1975).

Of the ERG potentials, the b-wave has been studied most extensively. The reasons for this are twofold. First, under most recording conditions, it is the most prominent ERG component and the most readily recorded. Second, it is the most sensitive of the ERG potentials. As the light stimulus is dimmed, the b-wave is the last to disappear in almost all cases. Figure 6.3 shows ERGs elicited from the skate eye over a range of 6 log units. With low stimulus intensities (log *I* = −6 and −4), only the b-wave is seen. The a-wave is evident only when the light intensity is raised more than 2 log units above b-wave threshold (that is, at log *I* = −2); and the initial portion of an obvious c-wave is observed in this recording only with the brightest stimulus. In the rat Cone (1963) has shown, using conventional recording techniques, that a measurable b-wave can be elicited with lights so dim that only 1 rod in 200 is absorbing a quantum. Thus, b-wave threshold is less than tenfold higher than the absolute human visual threshold and is comparable to the perceptual threshold.

The b-Wave and Müller Cells

It has long been recognized that the b-wave of the ERG arises proximal to the photoreceptors within the retina (Granit, 1947). It is blocked by agents that do not alter photoreceptor responses but do interfere with synaptic transmission between the photoreceptor cells and second-order elements. (See Figure 6.11, which shows the loss of the b-wave when high levels of aspartate are applied to the retina.) On the other hand, a perfectly normal ERG is recorded from human patients or animals whose ganglion cells have degenerated (Noell, 1953). Thus, ganglion cells, like the photoreceptors, do not appear to contribute directly to the generation of the b-wave. Granit (1947) proposed that the b-wave is initiated by membrane depolarization. He based his conclusion on the sensitivity of the b-wave to raised extracellular levels of KCl; and this idea has subsequently received support from a variety of studies (Brindley, 1960; Tomita, 1963; Byzov, 1965).

The notion that the b-wave has its origin in the Müller cells of the retina arose from two independent studies that were carried out in the late 1960s and used very different approaches. Faber under-

6.4 Comparison of an ERG recorded from the mudpuppy eyecup and an intracellularly recorded Müller cell response, also from the mudpuppy. The retina was evenly stimulated with a 1-sec light flash. The waveforms of the extracellularly recorded b-wave and the intracellularly recorded Müller cell response are very similar, as are the latencies of the two responses. Modified from Dowling (1970), with permission of J. B. Lippincott.

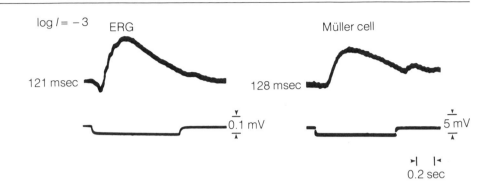

took an analysis of the extracellular potentials evoked intraretinally following a light flash by carrying out a second spatial derivative analysis of the voltage profiles corrected for variation in retinal resistance. From these data Faber was able to plot the "sources" and "sinks" of the extracellular b-wave current in the rabbit retina, that is, the retinal loci where current flows out of cells (sources) or into cells (sinks). He concluded that the main sink for the b-wave current lies in the outer plexiform layer, but that the current sources are both distal and proximal to that site. An important observation was that the proximal source extended all the way to the inner margin of the retina. These data indicated that the cells giving rise to the b-wave currents must extend from beyond the outer plexiform layer to the inner border of the retina. The only cells spanning that much of the retina are the Müller cells (see Figures 2.2 and 6.6), and Faber concluded that "the only retinal element which has a spatial distribution consistent with the distribution of b-wave sources and sinks is the Müller cell." Unfortunately, Faber's study was never published. This work formed part of his doctoral thesis, carried out under the supervision of W. Noell, and is available only in that form (Faber, 1969).

The second study was carried out by Miller and Dowling (1970), who sought to identify those intracellularly recorded responses in the mudpuppy that could be related to the extracellular b-wave. None of the responses recorded from the retinal neurons matched the properties of the b-wave; however, recordings made from Müller cells showed relatively slow, depolarizing potentials that were remarkably similar in the best cases to b-waves recorded from the retina. The potentials recorded from the Müller cells ranged in amplitude from 2 to 15 mV. The resting potentials of the cells were

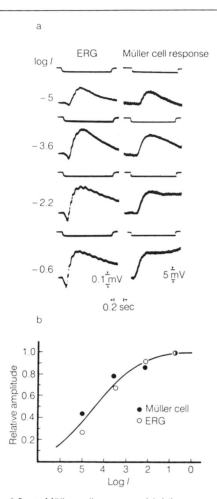

6.5 a Müller cell response (*right*) compared with the ERG (*left*) recorded over a range of flash intensities in the mudpuppy. At low light intensities, the b-wave and Müller cell responses show a very similar waveform. At higher intensities, the ERG is more complex due to an increased a-wave response and oscillatory potentials that are superimposed on the b-wave. Also, at bright flash intensities, the Müller cell response is of longer duration than is the b-wave.

b Intensity–response curves for the initial peak of Müller cell and b-wave responses. Over the five log units of intensity tested, the peak amplitudes were similarly related. Modified from Miller and Dowling (1970), with permission of the American Physiological Society.

variable, but some were as large as −85 mV, a level far beyond that ever recorded from mudpuppy neurons, but one typical for glial cells (Kuffler and Nicholls, 1966). In some instances the Müller cells were stained intracellularly, and dye was observed to extend from the inner margin of the retina to beyond the outer plexiform layer, the full extent of the Müller cells in the mudpuppy.

Figure 6.4 shows an intracellular response recorded from a Müller cell and an ERG elicited from the mudpuppy retina with the same light stimulus. The Müller cell response and the b-wave of the ERG are nearly identical in latency, and the waveforms of the two responses are quite similar. An a-wave and oscillatory potentials are observed in the ERG but are absent from the Müller cell recording, as would be expected, because these potentials appear to arise from different sources. Finally, a small off-response is observed in both the ERG and the Müller cell response.

When the Müller cell response and the b-wave are recorded over a range of stimulus intensities (Figure 6.5a) and compared, the latencies for both can be seen to correspond closely at all intensities. The waveforms of the two responses are quite similar at low stimulus intensities; but with brighter flashes, the Müller cell response is longer lasting. In addition, the voltage-intensity curves for the two responses correspond reasonably well (Figure 6.5b), and both curves are very broad, with a dynamic range of close to 5 log units. This range is very much greater than the range of V–log I curves found for any of the retinal neurons in the mudpuppy and provides additional evidence that the b-wave does not arise directly from one of the retinal neurons.

Figure 6.6 shows a schematic drawing of a Müller cell and Figure 6.7 shows several electron micrographs of various areas of the cell. A number of the anatomical features of the cell correlate well with the physiological findings. Intracellular recordings from Müller cells are obtained from areas representing a wide range of retinal depths but most frequently are made proximally, in the region of the ganglion cells or in the middle of the inner nuclear layer. Figure 6.6 shows that these depths correspond to the expanded endfoot region and the enlarged nuclear region of the cell. The receptive field of the Müller cell is very large (~1–2 mm), a finding suggesting electrical coupling between cells. And, indeed, such close apposition (gap) junctions are observed between Müller cells at the level of the outer limiting membrane (OLM) (Figure 6.7b). Electron microscopic studies show further that the Müller cells and their processes fill much of the extracellular space in the retina, leaving

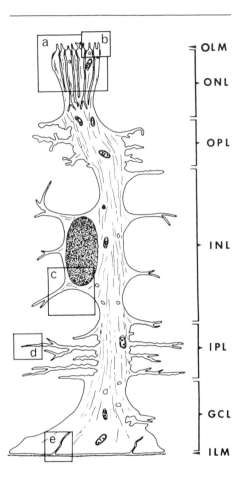

OLM
ONL
OPL
INL
IPL
GCL
ILM

6.6 Drawing of a Müller cell in the mudpuppy retina (based on light and electron micrographs). Its principal features are the numerous deep infoldings into the cell along its apical border, junctional complexes along the apical cell margin that form the outer limiting membrane, numerous processes that extend from the main column of cell cytoplasm to fit between and around the neuronal elements, and the expanded endfoot region of the cell. These features are illustrated in the electron micrographs of Figure 6.7, which show areas comparable to areas labeled a–e in this figure. Abbreviations defined in Figure 6.8. Modified from Miller and Dowling (1970), with permission of the American Physiological Society.

gaps only about 20 nm wide between the neural elements, and that they envelop, at least partially, all of the retinal neurons and many of their processes. The generation of the Müller cell responses undoubtedly relates to these close anatomical relationships.

Role of K^+

The experiments of Kuffler and his colleagues in the mid 1960s established that glial cells are permeable mainly to K^+ and that their large resting potentials (up to -90 mV) depend almost exclusively upon the large transmembrane gradient of K^+ (Kuffler and Nicholls, 1966). Changes in extracellular K^+ concentrations will affect membrane potential in accord with the Nernst equation; which relates potential to concentration gradients. Thus, the cells act like a K^+ electrode. Because the mudpuppy Müller cell responses appeared to be similar to the responses of glial cells recorded elsewhere (large resting potentials, etc.), Miller and Dowling (1970) proposed that the depolarizing Müller cell responses were the result of light-generated increases in K^+ concentration in the narrow extracellular spaces between the retinal neurons and the Müller cells. Experiments both earlier and later, showing the extreme sensitivity of the b-wave to extracellular K^+ levels, are consistent with this proposal (Granit, 1947; Miller, 1973).

Faber's analysis indicated a substantial sink for b-wave current in the region of the outer plexiform layer. Miller and Dowling (1970) suggested that K^+ increases occurring in the distal retina, and perhaps deriving from the horizontal or bipolar cells, underlie the Müller cell potentials. But subsequent measurements of K^+ levels at various depths of the retina following photic stimulation showed that the situation is considerably more complex than originally envisioned (Oakley and Green, 1976; Karwoski and Proenza, 1977, 1978; Kline et al., 1978; Dick and Miller, 1978). Three light-evoked changes in extracellular K^+ arising at three different intraretinal loci have been detected (Figure 6.8). In the region of the inner segments of the photoreceptors (distal to the outer limiting membrane), a large decrease in extracellular K^+ concentration occurs following a light flash, whereas two smaller increases in extracellular K^+ concentration occur deeper in the retina, one in the region of the outer plexiform layer (called the distal increase) and the other in the inner plexiform layer (called the proximal increase). The distal decrease of K^+ is unrelated to b-wave generation, but it does appear to underlie the generation of other components of the ERG (c-wave and the slow PIII). The increases in extracellular K^+

6.7 Electron micrographs of various regions of the Müller cell in the mudpuppy.

a Apical border of the cell showing the deep infoldings into the cell that occur at this level and the junctional complexes that form the outer limiting membrane (OLM). R, receptor.

b Higher power electron micrograph showing the junctions that are observed between the Müller cell processes and between the Müller cells (MC) and photoreceptors (R). Between the Müller cells and photoreceptors, only desmosomal-like junctions are usually seen (*left*). Between Müller cell processes, both desmosomal (*filled arrow*) and gap junction-like contacts (*open arrow*) are observed.

c Portions of two Müller cells (MC) in the inner nuclear layer. The nucleus (n) of one of the cells is in the upper left corner. The Müller cells and their processes fill most of the extracellular space between the neurons, leaving extracellular gaps of only 20–30 nm.

d On the left is part of a Müller cell extending through the inner plexiform layer. The arrow points to a fine process form an adjacent Müller cell. The open arrow points to a synapse.

e A part of the expanded endfoot region of the cell and the inner limiting membrane. Two infoldings into the cell are observed along the endfoot region of the cell. From Miller and Dowling (1970), reprinted with permission of the American Physiological Society.

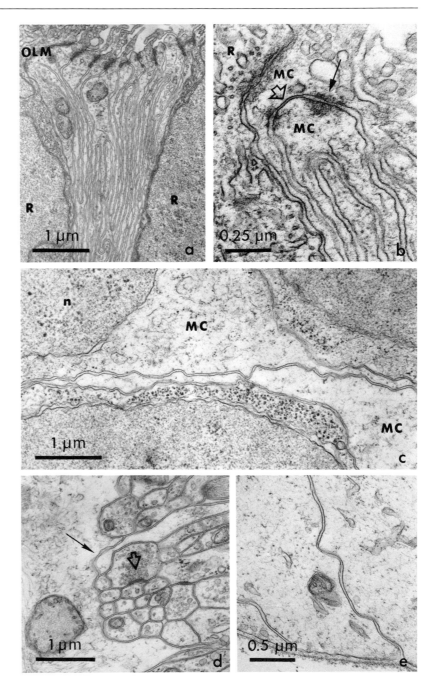

6.8 *Right,* K⁺ fluxes (KRGs) recorded at various depths within the skate retina and evoked with the same light intensity. Intraretinal K⁺ electrode position was estimated from readings on the micrometer drive used to push the electrode through the retina and are given relative to the distance between the inner limiting membrane (ILM) and the pigment epithelium.

Left, Schematic drawing of the skate retina. Relative retinal thickness for this drawing was based on histological sections of the skate retina. Distally in the retina (>60% retinal depth), a decrease in K⁺ concentration was observed in response to illumination. In the region of the outer plexiform layer (OPL), a sharp increase of K⁺ was observed following light stimulation, whereas proximally in the retina, in the region of the inner plexiform layer (IPL), a second, slower K⁺ increase was recorded. The arrow above the trace recorded at 44% retinal depth points to the off component of the response. OLM, outer limiting membrane; r, inner segment of the photoreceptor; ONL, outer nuclear layer; OPL, outer plexiform layer; h, horizontal cell; INL, inner nuclear layer; b, bipolar cell; a, amacrine cell; IPL, inner plexiform layer; g, ganglion cell; GCL, ganglion cell layer; ILM, inner limiting membrane. Modified from Kline et al. (1978), with permission.

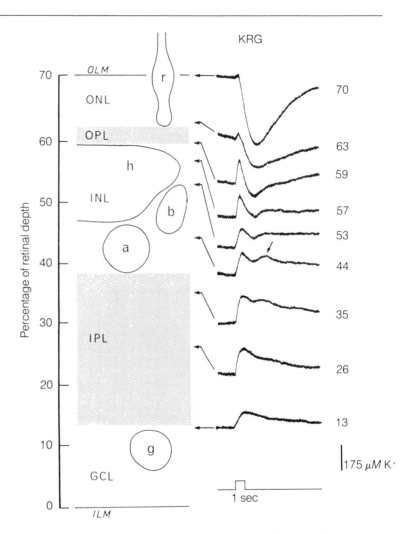

levels, on the other hand, both appear to contribute to the generation of Müller cell potentials (Karwoski and Proenza, 1977), but it is likely that the distal increase in K⁺ is mainly responsible for b-wave generation (Dick and Miller, 1978; Kline et al., 1978; Newman, 1980).

The distal increase in extracellular K⁺ following light stimulation appears to result from depolarizing bipolar cell activity (Dick and Miller, 1978). Following retinal illumination, this K⁺ efflux occurs earlier and is more transient than is the proximal K⁺ increase (Kline et al., 1978, 1985). The proximal K⁺ increase, on the other hand, appears to derive primarily from the activity of amacrine cells of the retina (Karwoski and Proenza, 1978). It is a slower

6.9 Intensity–response curves for the distal K⁺ increase and the b-wave, recorded simultaneously from the skate retina. Except at the brightest intensities, the two responses follow a parallel course over the intensity range tested. From Kline et al. (1985), reprinted with permission of Pergamon Journals Ltd.

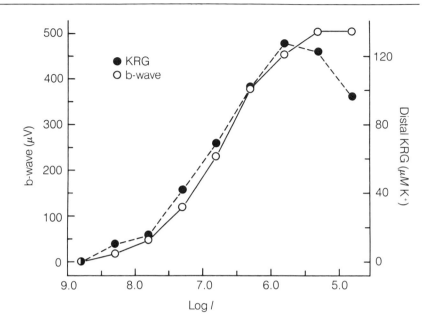

response than is the distal increase and exhibits, usually, on and off components (Figure 6.8). Evidence that the distal K⁺ increase is the major contributor to the b-wave currents is primarily threefold. First, the simultaneous application of ethanol and GABA to the frog retina enhances the distal K⁺ increase and reduces the proximal increase. These changes are associated with a marked increase in b-wave amplitude (Dick and Miller, 1978). Second, there is a close correspondence between the b-wave amplitude and increases in distal K⁺ levels, when related to light intensity (Kline et al., 1985). This correspondence is shown in Figure 6.9. And finally, the endfoot region of the Müller cell is more permeable to K⁺ than is most of the rest of the cell (Newman, 1984). Indeed, approximately 94 percent of the K⁺ permeability in the amphibian is confined to this region of the cell. This finding indicates that during b-wave generation K⁺ will flow out preferentially from that part of the cell; that is, the endfoot region is the major cellular exit point (current source) for K⁺. Thus, the distal K⁺ increase, because it generates a longer transretinal current loop (see Figure 6.10), will contribute substantially more to extracellular current flow (that is, to the b-wave) than will the proximal K⁺ increase.

Figure 6.10 is a diagram showing how the b-wave may be generated in the retina (modified from Newman, 1986). Results of a

6.10 *Left,* Currents giving rise to the b-wave. *Right,* A summary scheme of how the b-wave may be generated by the Müller cells (M) of the retina.

There are two current sinks and one major current source that underlie the b-wave. It is proposed that the two light-evoked increases of K⁺ that occur in the retina result in a flow of K^+ into the cell at those two loci (that is, the current sinks), while K^+ flows out from the cell at its endfoot region (that is, the current source). The extracellular current flow thus established around the Müller cell gives rise to the b-wave of the ERG. OPL, outer plexiform layer; IPL, inner plexiform layer. Modified from Newman (1980), with permission of the American Physiological Society.

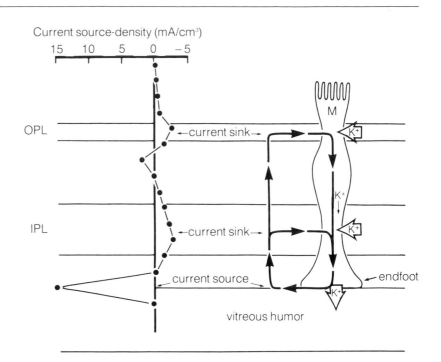

Current source-density (mA/cm³)

current density–source analysis are shown on the left and demonstrate that the b-wave is generated by two current sinks, one in the outer plexiform layer and a second in the inner plexiform layer, and a current source along the inner margin of the retina. The light-evoked increases in K^+ in the two plexiform layers cause K^+ influxes into and depolarization of the Müller cells. Almost all of the K^+ current exits from the high-conductance endfoot region of the cell. The extracellular current flow thus established around the Müller cells, particularly by the distal K^+ increase, gives rise to the b-wave of the ERG.

The scheme shown in Figure 6.10 is likely to be an oversimplification. Evidence obtained from the skate retina suggests that the b-wave is shaped by subtractive as well as by additive current flows. Thus, Kline et al. (1978) suggested that under some conditions the proximal K^+ increase might induce in the retina a current flowing in the direction opposite to that of the current established by the distal K^+ increase. Because the proximal K^+ increase is slower and longer lasting than the distal K^+ increase, this subtractive current flow would tend to make the b-wave more transient. This could

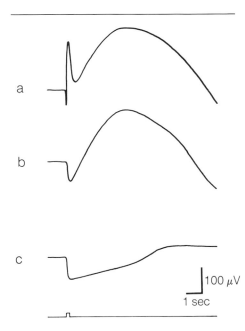

6.11 Isolation of the PIII component from the ERG of the skate retina.

a This trace was recorded from the untreated eyecup preparation and shows the a-, b-, and c-waves of the normal ERG.

b After immersion of the eyecup for a few minutes in Ringer solution containing aspartate, the b-wave of the ERG was suppressed, but the a- and c-waves were essentially unaltered.

c Removing the retina from the eyecup eliminated the c-wave, and only a corneal negative potential remained. Modified from Dowling and Ripps (1970), with permission of the Rockefeller University Press.

explain, for example, why the b-wave is more transient at higher stimulus intensities than is the Müller cell response (Figure 6.5). Some evidence in favor of this notion comes from the experiments of Dick and Miller (1978), who found that following suppression of the proximal K^+ increase by pharmacological means, the b-wave was somewhat prolonged.

The a-Wave Complex

The a-wave of the ERG can be readily isolated from the b- and c-waves. Figure 6.11 shows such an experiment on the skate eye. Initially, the eye was opened and the entire ERG recorded. The b-wave was then suppressed by flooding the eyecup for a few minutes with Ringer solution containing 100 mM sodium aspartate, which blocks transmission between the photoreceptors and second-order cells (Dowling and Ripps, 1972). Following this treatment, only the a- and c-waves were recorded. Finally, the retina was lifted free from the eyecup and pigment epithelium and placed between the recording electrodes. Subsequent recordings showed only corneal-negative potentials, that is, Granit's PIII component.

As already noted, the a-wave and the PIII component consist of several subcomponents. Indeed, three different components have been resolved in studies from various animals. Two arise from the photoreceptor cells and have been called the early and late receptor potentials—frequently abbreviated as ERP and RP, respectively. The third is a potential arising from the Müller cells and is usually called the slow PIII. The early receptor potential (ERP) is observed only when the retina is illuminated with very bright flashes; the late receptor potential (RP) and the slow PIII, on the other hand, are evoked with light flashes of ordinary intensity, and separating them is more of a problem. An effective way to do this is by placing microelectrodes across the receptors and recording differentially between them. This procedure limits the recording to currents generated mainly by the photoreceptors.

Figure 6.12 shows schematically how the late receptor potential and the slow PIII were separated in the carp retina (Witkovsky et al., 1973; see also Murakami and Kaneko, 1966). Recording across the retina (C–A) yielded a potential consisting of both the late receptor potential and the slow PIII. Recording across the receptor layer (B–A) yielded a rather square potential with fast onset and offset, that is, the late receptor potential; recording across the proximal retina (C–B) yielded a potential that developed and decayed slowly, that is, the slow PIII. Usually when recording across the

6.12 Schematic drawing showing the separation of the receptor potential and the slow PIII response in the isolated carp retina. One micropipette (A) was placed on the receptor surface of an isolated retina and another (B) was located at the level of the receptor terminal. A third electrode (C) was placed on the ganglion cell side of the retina. The b-wave was suppressed with aspartate and recordings were made across the whole retina (C–A), across the receptors only (B–A), and across the inner layers of the retina (C–B). The square potential recorded across the photoreceptor layer (B–A) is the photoreceptor potential; that across the inner layers of the retina (C–B) is slow PIII. INL, inner nuclear layer; GCL, ganglion cell layer. After Witkovsky et al. (1973), with permission of the Rockefeller University Press.

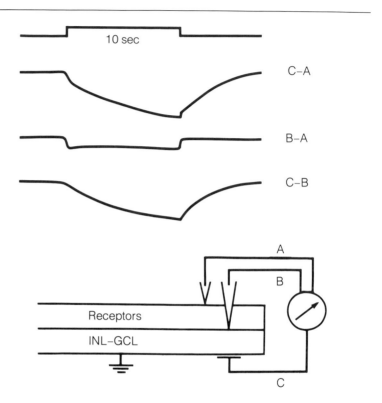

whole retina in an experiment such as that shown in Figure 6.12, the leading edge of the receptor potential can be distinguished from the slow PIII by virtue of its faster onset.

That the isolated late receptor potential, and the leading edge of the a-wave, represent photoreceptor activity has been shown in a number of experiments (Brown, 1968; Penn and Hagins, 1969). For example, in the mudpuppy the a-wave and the intracellularly recorded receptor response have a very similar latency, and the leading edges of the two types of responses have very similar shapes (Miller and Dowling, 1970). Furthermore, the voltage–intensity function of the late receptor potential is identical with that evoked from photoreceptors; that is, it conforms with the Michaelis-Menten equation with a slope of 1 (Hagins et al., 1970).

The slow PIII, in contrast to the late receptor potential, rises in amplitude only very slowly. In the carp, increases in amplitude occur for the duration of a 30-sec stimulus; in other words, the potential integrates the effects of prolonged illumination over time (Witkovsky et al., 1975). When a voltage–intensity function is plotted for the carp slow PIII evoked with long-duration flashes, however, the relationship is identical to that seen for rod photoreceptors. That is, it conforms to the Michaelis-Menten relationship with a slope of 1 and it requires about 30 quanta to half-saturate the response.

Furthermore, the spectral sensitivity function for the slow PIII in the carp conforms closely to the absorption spectrum of the rod photopigment. These data indicate that the slow PIII response in the carp closely reflects rod receptor activity; but the fact that the potential is recorded across the proximal retina indicates that its origin is in another structure (Witkovsky et al., 1975). Current source–density analysis of the slow PIII in the rabbit indicated that the source of the potential is located in the region of the inner segments of the photoreceptor cells, with current sinks extending to the inner margin of the retina (Faber, 1969). The Müller cell is the only retinal element extending this distance in the retina, and it has been postulated that, like the b-wave, the slow PIII arises from currents generated around the Müller cell.

The observation that there is a large decrease in K^+ levels in the vicinity of the inner segments of the photoreceptors (Figure 6.8) provides a suggestion as to how the slow PIII response arises in the Müller cell. The decrease of K^+ in the extracellular space around the inner segments of the photoreceptors will induce K^+ to flow out from the distal region of the Müller cell and into the proximal, endfoot region of the cell from the vitreous humor; the resulting proximally directed extracellular current flow along the length of the Müller cell gives rise to the corneal-negative slow PIII potential.

Early Receptor Potential

The early receptor potential, as already noted, is observed only when a retina or an eye is illuminated with a very intense, short (~1 msec) flash (Brown and Murakami, 1964; Cone, 1964). In such experiments, shown schematically in Figure 6.13, two waves can be detected preceding the a-wave. The first, smaller potential, called R_1, is corneal positive and arises without detectable latency as the flash is presented. The second wave, called R_2, is corneal negative and is interrupted by the a-wave in a recording from an intact retina or eye.

The origin of these two photovoltages has been deduced by two kinds of experiments. First, these fast potentials remain after all other ERG and neuronal potentials disappear following drastic treatments of the eye or retina; for example, anoxia, cooling,

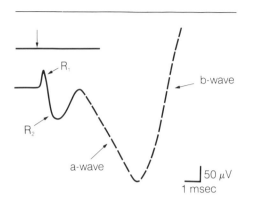

6.13 Schematic drawing illustrating the R_1 and R_2 components of the ERP and their relation to the a- and b-waves of the ERG. This drawing is based on recordings made in the rat (Cone, 1964). A short (1-msec), intense flash of light presented to the retina results in the generation first of a corneal-positive potential (R_1), following by a corneal-negative potential (R_2). Both of these potentials occur before the a-wave begins. The arrow at the upper left indicates flash onset.

altered ionic and pH levels, and partial fixation (Brown and Murakami, 1964; Cone, 1964; Pak, 1965; Brindley and Gardner-Medwin, 1966). As long as the outer segments of the photoreceptors remain intact, an ERP can be evoked. Second, to record an ERP requires a light flash sufficiently bright to excite a significant number of visual pigment molecules (>10 percent). Furthermore, the amplitude of the response is related directly to the amount of visual pigment excited by the flash: if most of the visual pigment has been inactivated (bleached), no ERP can be evoked, but once visual pigment is restored, an ERP can again be recorded (Cone, 1967).

These experiments indicate that the ERP arises from the outer segments of the photoreceptor cells and that this potential is not generated in the same way neuronal or glial potentials are. Rather it appears to arise from the visual pigment molecules themselves, which undergo conformational changes as a result of light absorption. These changes cause a charge displacement within the molecules. There are large numbers of these molecules in each photoreceptor outer segment ($\sim 10^9$) and large numbers of photoreceptors in a retina ($\sim 10^8$); and the visual pigment molecules are highly oriented in the membranous disks of the outer segment. Consequently, the virtually simultaneous excitation of a significant number of molecules by a bright flash results in the generation of a current flow around the outer segments that can be detected across the retina and even across the eye.

It is clear that the ERP results directly from changes induced in the visual pigment molecules by light. Thus, it provides a way to probe the nature and dynamics of the visual pigment excitation and bleaching process that occurs in the photoreceptors in intact or living photoreceptors (Cone, 1967; Cone and Cobbs, 1969; Goldstein and Berson, 1969). This topic will be discussed in the next chapter.

The c-Wave and Pigment Epithelium

The experiment illustrated in Figure 6.11 showed that a c-wave is not recorded from an isolated retina. This observation clearly indicates that the c-wave does not arise from within the retina itself. The idea that the c-wave is generated by the pigment epithelium was suggested first in the 1950s by Noell, who observed that the c-wave disappears after the pigment epithelium is selectively destroyed by the metabolic poison iodate (Noell, 1953, 1954). Noell's suggestion has since been supported by a variety of studies, and

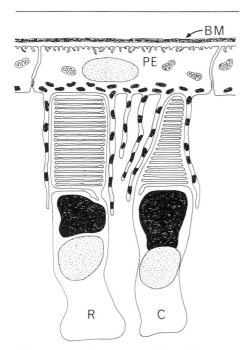

6.14 A schematic drawing of the relation between the pigment epithelial (PE) cells and a rod (R) and a cone (C) in the mudpuppy retina. Villous processes containing pigment granules extend from the pigment epithelial cells and lie along the outer segments of both kinds of receptors. Adjacent PE cells form close appositional junctions, which serve to seal the extracellular space around the receptors from that found in the rest of the back of the eye. Some infoldings are found along the distal margin of the cell, and above those infoldings is a basement membrane (BM).

compelling evidence has been provided that the decrease in extracellular levels of K^+ that occurs around the photoreceptors when the retina is illuminated leads to c-wave generation. Thus, the c-wave, the slow PIII, and the b-wave all appear to result from the altered K^+ fluxes in the retina induced by light.

The pigment epithelium lines the back of the eyecup and is in intimate contact with the photoreceptors, primarily via long villous processes that extend between the outer segments and partially ensheath them (Cohen, 1963; Steinberg and Wood, 1974). Figure 6.14 illustrates the relation of the pigment epithelial cells and rod and cone outer segments (see also Figure 2.12). This drawing is based on observations made in the mudpuppy, and there, as in many species, the pigment epithelial cell processes extend down to the level of the inner segments. Pigment granules are observed in the villous processes; and in cold-blooded vertebrates (but not mammals) these pigment granules move during light and dark adaptation. In the dark the granules are clustered in or close to the cell perikaryon; but in the light the pigment granules move along the villous processes to lie between the outer segments, partially shielding them. (In the light-adapted retina, the presence of numerous pigment granules in the villous processes between the outer segments holds the retina tightly to the back of the eye. Thus, it is very difficult to remove the retina intact from the eyecup of a light-adapted, cold-blooded vertebrate such as a frog.) Adjacent pigment epithelial cells form close appositional contacts along the apical border of the cells. These contacts appear to be both tight and gap junctions. The tight junctions form a seal between adjacent pigment epithelial cells, isolating the extracellular space around the photoreceptors from the extracellular space between the pigment epithelial cells and the rest of the back of the eye. The gap junctions, on the other hand, presumably allow for electrical continuity between the pigment epithelial cells.

Intracellular recordings from pigment epithelial cells of both cats and frogs have been made, and in both instances the intracellularly recorded potentials and the extracellularly recorded c-wave were correlated (Steinberg et al., 1970; Oakley et al., 1977). Figure 6.15 shows these potentials and a record of the K^+ levels in the distal region of the frog retina. Resting potentials of the pigment epithelial cells averaged -83 mV in the frog; and in the experiment illustrated in Figure 6.15, a hyperpolarizing potential of about 10 mV was recorded from the cell. In the cat, responses of up to 20 mV have been recorded from the pigment epithelium (Steinberg et al.,

6.15 Light-evoked responses recorded from the frog retina. The top trace is the ERG, showing a-, b-, c-, and d-waves; the middle trace is an intracellular recording of the membrane potential of a pigment epithelial cell; and the bottom trace is a KRG record of K⁺ levels just proximal to the pigment epithelial cells, in the region of the photoreceptor inner and outer segments. The stimulus (*bottom line*) was a 50-sec light flash. There is excellent correspondence between the c-wave of the ERG, the intracellularly recorded response of the pigment epithelial cell, and the change in K⁺ levels. Modified from Oakley et al. (1977), with permission of J. B. Lippincott Company.

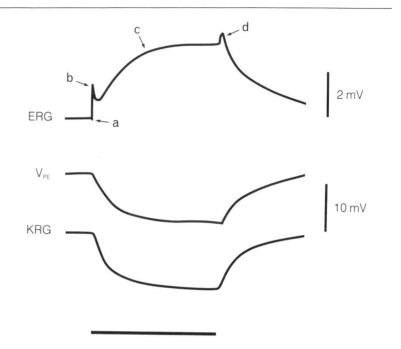

1970). Furthermore, pigment epithelial cells are very sensitive to changes in extracellular K⁺ concentrations and, like Müller cells, behave like K⁺ electrodes. As can be seen in Figure 6.15, there is an excellent correspondence between the waveform of the intracellularly recorded pigment epithelium potential, the c-wave, and the changes in K⁺ around the photoreceptors.

As noted earlier, the c-wave is a prominent feature of ERGs recorded from rod-rich retinas but not (usually) from cone-dominated retinas. In rod-dominated retinas the slow c-wave appears to be generated solely by rod activity (Oakley and Green, 1976). In the cone-dominated turtle retina both hyperpolarizing potentials in the pigment epithelial cells and decreases in K⁺ around the photoreceptors have been evoked in response to light (Matsuura et al., 1978). These responses are much faster than those seen in rod-dominated retinas—and considerably smaller. Thus, an extracellular c-wave potential appears to be generated in cone-dominated eyes even though it is normally swamped by the other ERG waves. It is possible, however, to unmask a c-wave in the turtle by depressing the b-wave potential (Matsuura et al., 1978).

A final question is, What leads to the large decrease in K⁺ around the photoreceptors during illumination? A model was suggested by

Matsuura et al. (1978) and has received experimental support from Oakley et al. (1979). As noted earlier, the outer segments of the photoreceptors are permeable to Na^+ in the dark; that is, Na^+ is continually flowing into the outer segment. This inward flux of Na^+ is balanced by an outward flow of K^+ from the inner segment region of the cell. These fluxes give rise to a dark current that flows around the photoreceptors (Penn and Hagins, 1969; Hagins et al., 1970). To maintain the appropriate ionic concentrations within the photoreceptor, a powerful Na^+/K^+ exchange pump, located in the inner segment membrane, extrudes Na^+ from the cell and pumps K^+ back in. In the dark the passive efflux of K^+ from the photoreceptor is balanced by an equal influx of K^+ from the Na^+/K^+ pump. In the light Na^+ permeability of the outer segment is rapidly shut down, and the passive efflux of K^+ from the inner segment is thereby decreased. The Na^+/K^+ exchange pump, however, is not turned off by illumination. Therefore, it reduces the extracellular levels of K^+ in the vicinity of the inner segments. It is this decrease in K^+ levels around the photoreceptors that is believed to give rise to the c-wave (and also to the slow PIII).

The d-Wave

The d-wave, or off-response of the ERG, is seen prominently in recordings from cone-dominated retinas but has not been analyzed nearly as extensively as the a-, b-, and c-waves. Off-responses are observed in Müller cell recordings (Miller and Dowling, 1970); thus, the Müller cell may underlie the generation of at least part of the d-wave potential. On the other hand, it has long been supposed that the rapid turn-off of the cone photoreceptor response contributes substantially to the d-wave (Granit, 1955); and in the mammalian eye there is evidence to support this notion (Brown, 1968).

Minor Components: The Oscillatory Potentials

As shown in Figures 6.4 and 6.5, small wavelets superimposed on the b-wave of the ERG are frequently observed (Cobb and Morton, 1952; Yonemura et al., 1963; Ogden, 1973). These wavelets, called oscillatory potentials, are not seen in Müller cell recordings, a finding suggesting that these potentials are distinct from the b-wave. Indeed, a number of studies have demonstrated the independence of the oscillatory potentials from the b-wave (Nye, 1968; Wachtmeister, 1972).

The oscillatory potentials are most easily observed in the partially light-adapted eye, because they have a higher light threshold

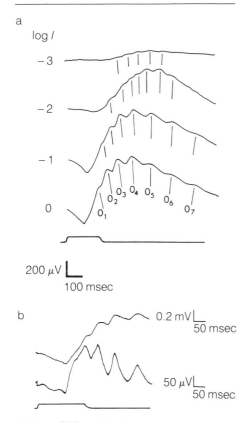

a

log *I*

-3

-2

-1

0

O_1 O_2 O_3 O_4 O_5 O_6 O_7

200 μV

100 msec

b

0.2 mV
50 msec

50 μV
50 msec

6.16 **a** ERGs elicited from the mudpuppy eyecup preparation by stimuli of various intensities delivered with a constant interstimulus interval (30 sec). Each trace is a typical record elicited in response to repetitive stimuli. At all intensities, oscillatory potentials can be observed; at the highest stimulus intensity employed (log *I* = 0), seven oscillatory potentials (O_1–O_7) can be distinguished.

b ERGs elicited with a bright stimulus (log *I* = 0) repeated at 30-sec intervals and recorded at a rate twice that used for the records shown in a. The upper trace shows a record obtained with a long time constant amplifier setting (1 sec); the lower trace with a short time constant setting (20 msec). From Wachtmeister and Dowling (1978), reprinted with permission of J. B. Lippincott Company.

than the b-wave. Thus, an effective way to elicit large-amplitude oscillatory potentials is to use repetitive, relatively bright, full-field flashes with a constant interstimulus interval. In the mudpuppy an interstimulus interval of 60 sec was found to be optimal for eliciting the oscillatory potentials, whereas in humans an interstimulus interval of 30 sec was best (Wachtmeister, 1972; Wachtmeister and Dowling, 1978). In both the mudpuppy and the human ERG, up to seven oscillatory potentials are consistently seen, and these wavelets are reproducible for a particular set of stimulus conditions. Figure 6.16a shows recordings from the mudpuppy eyecup at four different stimulus intensities. At all but the dimmest stimulus intensity, seven oscillatory potentials (O_1–O_7) could be discerned. By reducing the time constant of the amplifier, the oscillatory potentials can be made much more obvious (Figure 6.16b); and as long as the time constant is not made too short, the peak times of the potentials are not significantly affected by this procedure.

The origins of the oscillatory potentials are not known. However, there is evidence that various oscillatory potentials arise at different levels within the retina. Some of the oscillatory potentials reverse polarity at different depths when they are recorded with a micropipette passing through the retina. The first oscillatory potential, O_1 reverses first, at about 20 percent of the retinal depth; O_2 and O_3 reverse next, at about 30 percent of the retinal depth; and O_4 and O_5 reverse even more proximally, at nearly 40 percent of the retinal depth. (No reliable measures of the reversal points of O_6 and O_7 have been made.) The notion that various oscillatory potentials arise from different sources is supported also by the observations that certain of the oscillatory potentials were enhanced significantly by changing interstimulus intervals, whereas others were not, and that certain of the oscillatory potentials appeared more sensitive to the application of GABA, glycine, and glutamate to the retina than were others (Wachtmeister and Dowling, 1978).

The finding that the earlier oscillatory potentials reverse polarity more proximally in the retina than do the later ones (that is, O_1 reverses before O_2 and O_3, which reverse before O_4 and O_5) suggests that the chain of events underlying these potentials starts in the proximal part of the retina and travels distally. One possible explanation for several extracellular current loops moving from the proximal to the distal retina would be a series of feedback interactions initiated in the proximal retina (Brown, 1968). Two possible feedback circuits that might underlie the oscillatory potentials are, first, the feedback from amacrine cell processes onto bipolar cell terminals and other amacrine cell processes in the inner plexiform layer and, second, the feedback from the interplexiform cells onto the horizontal and bipolar cells in the inner nuclear and outer plexiform layers.

Other Field Potentials: The Proximal Negative Response

When a low-resistance micropipette is inserted into the retina and extracellular recordings made at different retinal depths, the potentials observed can almost always be related to the various components of the ERG (Brown and Wiesel, 1961; Brown, 1968). However, when the micropipette or electrode is positioned at the border of the inner nuclear and inner plexiform layers and the retina stimulated with a spot of light, a potential is evoked that is not seen when recordings are made across the retina—that is, this response is distinct from the ERG. The potential is always negative and very transient in nature when evoked with suprathreshold stimuli; it has been called the proximal negative response (PNR) (Burkhardt, 1970).

Figure 6.17 shows PNR recordings from the retina of a dark-adapted skate (Dowling and Ripps, 1977). Near threshold (log $I = -8$), the response had a relatively long latency and consisted of a number of small, transient potentials. When evoked with brighter stimuli, the latency of the response shortened dramatically and a sizable on-transient dominated the response. With the brightest stimulus (log $I = -5$), sharp, spikelike potentials were seen riding on the leading edge of the on-transient. In other retinas the PNR shows both on and off components (Burkhardt, 1970; Proenza and Burkhardt, 1973). In the skate an off component is not observed when the retina is dark-adapted, but during light adaptation a prominent off-transient appears (Figure 6.17c).

The PNR has a waveform similar to that of the transient amacrine cells, and several pieces of evidence support the notion that this extracellular potential reflects transient amacrine cell activity. First, the level of maximum PNR recording, at the border of the inner nuclear and inner plexiform layers, is appropriate for a response arising from amacrine cells. Second, intracellular responses from transient amacrine cells resemble closely the PNR response in various species. For example, in the dark-adapted skate retina, most transient amacrine cells show only an on-transient (Figure 6.17b), which closely resembles the on-transient of the PNR. Fi-

6.17 a Proximal negative responses (PNRs) recorded from a dark-adapted skate eyecup preparation over a range of stimulus intensities. Near threshold (log *I* = −8), the response consists of a slow depolarization, on which are superimposed a number of small, transient potentials. At brighter flash intensities, a prominent on-transient characterizes the response. Often one or a few spikelike potentials are observed riding on the on-transient.

b An intracellularly recorded response from the skate retina, presumably from an amacrine cell.

c A PNR recorded from a skate eyecup preparation that was partially light-adapted and shows a prominent off-response. Records from Dowling and Ripps (1977), reprinted with permission of the Rockefeller University Press.

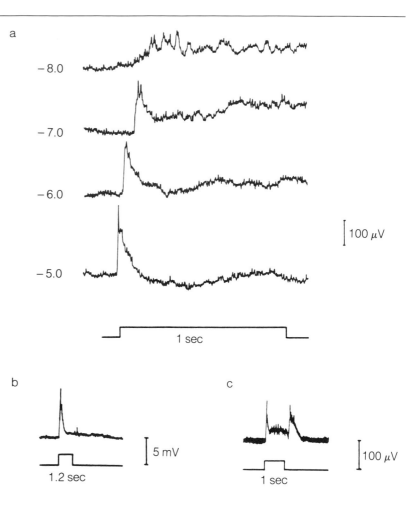

nally, and perhaps most compelling, the voltage–intensity relationship of the PNR is very steep and matches closely the *V*–log *I* function of amacrine cells. This relationship is illustrated in Figure 6.18, which shows PNR data from the frog (Burkhardt, 1970) and amacrine cell responses obtained by intracellular recording in the mudpuppy (J. E. Dowling, unpublished observations).

The PNR has proved useful for the study of the quantitative relations between transient amacrine cell activity and the impulse discharge of single on–off ganglion cells (Burkhardt and Whittle, 1973). In the frog the on-response of the PNR and the on-discharge of the on–off cells were virtually coincident in terms of latency and

6.18 Intensity–response function for PNRs recorded in the frog retina (*geometric symbols*) and an intracellularly recorded amacrine cell response in the mudpuppy (*stars*). The *V*–log *I* curve is very steep for both the PNR and intracellularly recorded amacrine cell response. Modified from Burkhardt (1970), with permission of the American Physiological Society.

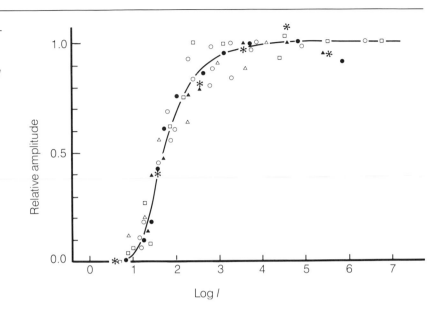

also in terms of normalized PNR amplitude and firing frequency of the ganglion cells. These data thus provide evidence that the on–off ganglion cells in the frog receive their excitatory input from the transient amacrine cells (Chapter 4). The PNR has also been useful in studies monitoring the effects of altered ionic enviroments on transient amacrine cell activity in the rabbit retina, where long-term intracellular amacrine cell recordings are difficult to make (Miller and Dacheux, 1973).

In summary, there is a rich variety of extracellular potentials that can be recorded from the retina and the eye. Some of the potentials arise directly from the neuronal cells (the PNR and the late receptor potential), but others appear to arise from the glial (Müller) cells or the pigment epithelium (b-wave, slow PIII, and c-wave). One extracellular potential (ERP) reflects molecular transitions occurring in the visual pigment molecules in the outer segments of the photoreceptors. And finally there are potentials (the oscillatory potentials) whose origins are unknown but that may reflect feedback circuits within the retina. The relationships between the various retinal cells, the extracellular potentials, and the responses these potentials reflect are summarized in Table 6.1.

An extracellular potential can be linked to each of the classes of

Table 6.1 Retinal cells and extracellular potentials (RP, late receptor potential; ERP, early receptor potential; PNR, proximal negative response)

Cell type	Potential generated	Precursor of response
Pigment epithelium	c-wave	Photoreceptor activity
Photoreceptors	a-wave, RP	Photoreceptor light response
Photoreceptors	ERP	Visual pigment transitions
Müller cells	b-wave	Depolarizing bipolar cell activity
Müller cells	Slow PIII	Photoreceptor activity
Amacrine cells	PNR	Transient amacrine cell responses

the intrinsic retinal cells and to the pigment epithelium, except for the horizontal cells. The horizontal cells, on the other hand, are the easiest of the retinal neurons to record from intracellularly, and we have much detailed information concerning the responses of these neurons (see Chapter 4).

Visual Adaptation and Photoreceptor Mechanisms

A STRIKING feature of the visual system is its ability to adapt to background illumination and to function over an enormous range of light intensities. The human eye can discriminate visually over a luminance span of about 10 billion to 1. The visual system accomplishes this by adjusting its light sensitivity relative to the ambient level of illumination. One obvious mechanism to explain light and dark adaptation is the regulation of the amount of light entering the eye by the pupil. In bright light the pupil closes down rapidly; but in the human eye the diameter of the pupil varies between 2 and 8 mm, a range allowing for only a 16-fold change in area. This mechanism can account for adaptation over a range of only about 1 log unit. Thus, the major part of adaptation must be attributable to mechanisms occurring within the retina or elsewhere in the visual pathway.

That the mechanisms underlying light and dark adaptation are largely retinal in origin has been appreciated for many years. Craik and Vernon showed in 1941 that light and dark adaptation proceeded perfectly normally in a human subject whose eye was pressure-blinded during the light adaptation period. In this experiment the adapting light was not seen, nor were central neurons stimulated during the light adaptation period. But when the pressure-blindness was relieved, visual sensitivity measured via that eye was decreased to the same extent as in a control, non-pressure-blinded eye, and the dark adaptation curve subsequently measured was perfectly normal. Furthermore, it has been shown in numerous experiments that the enucleated eye, or even the isolated retina, shows the full range of adaptive phenomena observed in intact animals or in humans.

In behavioral experiments with humans or animals the state of light and dark adaptation is usually evaluated by measuring the minimum intensity of light necessary to evoke a visual sensation, that is, a visual threshold (Crawford, 1937). Figure 7.1 illustrates the effect of increasing illuminance on visual threshold in a human subject (a typical light adaptation experiment) and the recovery of visual threshold following a bright-light adaptation period (a typi-

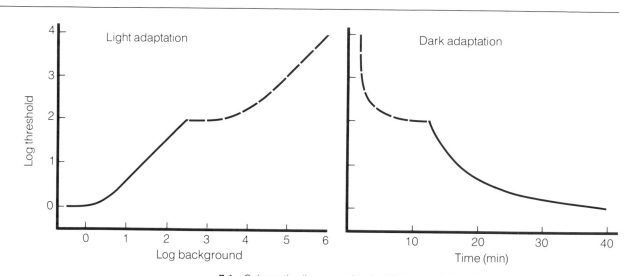

7.1 Schematic diagrams of typical light and dark adaptation experiments in the human. *Left,* Light adaptation: the rise of visual threshold is plotted as a function of background light intensity. *Right,* Dark adaptation: the recovery of visual threshold following bright-light adaptation is plotted as a function of time in the dark. In both kinds of experiment, two segments are observed. The upper segments (*dashed lines*) reflect cone system adaptation; the lower segments (*continuous lines*), rod system adaptation. See text.

cal dark adaptation experiment). When a background light is turned on, most of the loss of visual sensitivity, or much of the rise of visual threshold, occurs very rapidly—within a second or two; so what is typically plotted, as here, is the steady-state threshold level as a function of adapting intensity (Stiles, 1949). During dark adaptation, the recovery of sensitivity occurs relatively slowly— over many minutes; so what is typically plotted, as here, is the visual threshold as a function of time (Hecht et al., 1937).

As can be seen in Figure 7.1, both light and dark adaptation functions show two segments, which reflect adaptation of the rod and cone systems. The rod system governs dim light vision, so as background illumination is increased, the rod system loses sensitivity first (continuous line). When rod threshold is raised about 100-fold (2 log units), the cone system (dashed line) is now more sensitive and governs visual sensitivity. As the background intensity is raised further, the cone system also loses sensitivity and its threshold increases. For both the rod and cone systems, much of the increase in threshold is linearly proportional to the adapting luminance; in other words, the slope of the line relating threshold and background intensity is almost 1. This is the well-known Weber-Fechner relation in which $\Delta I/I = c$, where ΔI is the increment threshold, I is the background intensity, and c is a constant value of 0.1 when size and duration of the stimulus is optimally adjusted. Thresholds measured against background illumination are called increment thresholds.

Contributions from both rod and cone systems are seen during dark adaptation also. Initially, following bright-light adaptation, the cone system governs visual sensitivity. The cone-mediated system threshold falls relatively rapidly during dark adaptation and reaches a plateau within about 5 min. After approximately 10 min, the rod system takes over, and rod thresholds fall gradually over the next 20 min to reach maximum sensitivity, that is, the absolute threshold.

The data shown in Figure 7.1 suggest that both the rod and cone systems adapt to background illumination and that, although they show differences in absolute threshold and in the kinetics of threshold recovery, the underlying mechanisms of adaptation may be similar. A large number of experiments have indicated that this is generally correct. There are some differences in how the two systems adapt, and some species differences, but in general the mechanisms underlying light and dark adaptation appear to be similar in both systems. To begin to analyze the underlying mechanisms of visual adaptation, it makes sense, therefore, to turn to experimental situations where one system or the other predominates, or is the only system operating, and to retinas where adaptive mechanisms can be explored at several different levels, including within single cells.

Figure 7.2 shows an early attempt to do such an experiment using the rod-dominated eye of the rat (Dowling, 1963). At the time the experiment was done, it was thought that the rod system governed the entire adaptation range in this animal, because, using ordinary procedures, no cone limb is obvious in either the light or the dark adaptation curves. Green (1973) has since shown that the cones contribute to light-adapted thresholds in the rat, but this finding does not negate the points that this experiment demonstrates. Furthermore, results similar to those shown in Figure 7.2 have been obtained in the skate retina, which has no cones (Dowling and Ripps, 1970).

Since the formulation of Hecht's classic photochemical theory in the 1920s, one persisting idea has been that the changes in sensitivity of the retina during light and dark adaptation are related to the

7.2 The effects of 5 min of background light adaptation over a range of intensities on rhodopsin content (*filled circles, dotted line*), increment thresholds (*open circles, thick line*), and dark adaptation (*crosses, thin lines*). The data were obtained from eyes of anesthetized rats.

The rhodopsin content was not significantly reduced until the eye was exposed to background light intensities greater than log I = 4. The log increment threshold, on the other hand, rose linearly with intensities greater than log I = 0.5. Dark adaptation (*crosses*) was rapid until the eye was adapted to lights that bleached a significant fraction of the rhodopsin (that is, greater than log I = 4). Thereafter, dark adaptation consisted of two components, an initial rapid phase followed by a slower component. The extent of the slow component of dark adaptation depended on the amount of rhodopsin bleached. From Dowling (1963), reprinted with permission of the Rockefeller University Press.

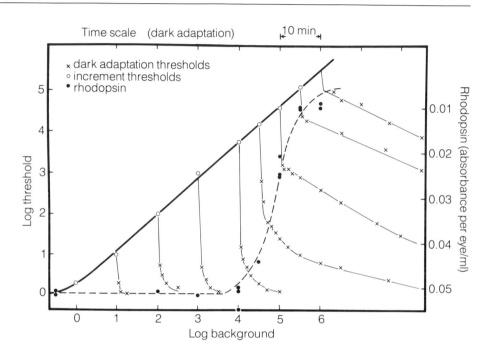

breakdown (bleaching) of visual pigment in the light (Plate 7) and regeneration of the visual pigment in the dark (Hecht, 1942). The primary evidence for this view came from observations that the time course of dark adaptation follows roughly the time course of visual pigment regeneration in the eye. However, numerous experiments, some dating back to the 1930s, indicated that factors other than breakdown and resynthesis of visual pigment must also be involved in visual adaptation. The evidence for this came from experiments that showed that there may be large changes in visual sensitivity without significant changes in the concentration of visual pigment (Granit et al., 1939; Rushton and Cohen, 1954).

The experiments shown in Figure 7.2 were designed to clarify the contributions of photochemical and nonphotochemical factors in light and dark adaptation. They combined in one set of data the kinds of experiments shown in Figure 7.1, and they added measurements of visual pigment (rhodopsin) levels remaining in the photoreceptors after exposure to the various adapting lights. The sensitivity of the rat eye was determined by measuring the intensity of light necessary to evoke a small, threshold ERG b-wave response, and the experiments were done in the following way. An eye was exposed to an adapting light of a given intensity and an increment threshold determined 4.5 min after the light was turned on. After 5 min, the adapting light was turned off and either the eye was immediately analyzed for its rhodopsin concentration or a dark adaptation curve was determined.

As can be seen in the figure, except at the lowest intensities the b-wave threshold rose linearly with adapting luminance; that is, the increment threshold followed the Weber-Fechner relation with a slope close to 1. The loss of b-wave sensitivity, however, appeared to be quite unrelated to changes in rhodopsin concentration. Indeed, no measurable breakdown (bleaching) of pigment occurred until the background light was 4–5 log units above the b-wave threshold. This is so because the eye responds to very low intensities of light (1 quantum per 200 rods to elicit a b-wave), but at the same time it contains large amounts of pigment (10^9 molecules/rod). At high adapting intensities that substantially bleached pigment, the incremental threshold still rose at about the same rate as it did with lower, nonbleaching intensities. Thus, the elevation of threshold during light adaptation appeared to depend almost entirely on the intensity of the background light, not on the amount of visual pigment bleached. It is true that if a substantial fraction of the pigment is bleached, fewer quanta can be caught and threshold will increase (see Figure 7.13). However, over almost all of the adaptive range this increase in threshold is small compared to the total change of threshold during light adaptation. For example, with half the pigment gone, the threshold would be doubled, but this is a change of only about 0.3 log units, compared to a total elevation of threshold of 4–5 log units induced by that adapting intensity. Only when virtually all of the visual pigment is bleached (>99 percent) will pigment bleaching be a significant factor in the rise of incremental threshold levels. Thus, the first conclusion to be drawn from these experiments is that loss of visual sensitivity during light adaptation is essentially unrelated to visual pigment loss.

Dark adaptation, on the other hand, is more complicated, and two distinct phases of recovery could be distinguished. Following exposure to dim adapting intensities that did not bleach measurable quantities of visual pigment, dark adaptation was very rapid and was usually completed within a minute or so. Following exposure to adapting intensities bright enough to bleach visual pigment, a slow phase of dark adaptation was also observed. The extent of the slow phase of adaptation was related to the amount of pigment bleached. When virtually all the pigment was bleached, the slow component accounted for almost all of dark adaptation, and complete threshold recovery required more than 90 min in the rats. Parallel measurements of the recovery (regeneration) of visual pigment and of the logarithm of the threshold during slow dark adaptation showed they were closely correlated (Dowling, 1963). In the human eye a similar relation between the recovery of log threshold and visual pigment regeneration holds for both rods and cones after they have been exposed to light bright enough to bleach a large fraction of the visual pigment (Rushton, 1961, 1965b).

Thus, during dark adaptation there is a slow component related to the level of visual pigment in the eye, called photochemical adaptation, and a faster component not apparently related to pigment concentration, often called neural adaptation (Dowling, 1963). Because the visual pigment resides in the outer segments of the photoreceptors, these experiments further indicated that during photochemical adaptation, sensitivity of the retina is governed by the photoreceptors.

These experiments immediately raise a number of questions. First, do the photoreceptors also govern light adaptation and the fast component of dark adaptation, or are there mechanisms that reside elsewhere in the retina that control visual sensitivity under these conditions? And second, what are the cellular mechanisms that underlie changes in photoreceptor sensitivity? That is, how does the bleaching of visual pigment change the light sensitivity of the photoreceptors so profoundly? We have been much more successful in answering the first question than the second. We can say with assurance that a good deal of visual adaptation is governed by the photoreceptors, but it is also clear that nonreceptoral adaptive mechanisms also exist and govern visual sensitivity under certain conditions. On the other hand, we can still say little with regard to the cellular mechanisms underlying visual adaptation, although we can describe in some detail the phenomena of adaptation in photoreceptors and other cells.

Photoreceptor Mechanisms

Figure 7.3 shows electron micrographs of portions of mudpuppy rod and cone outer segments (see Figure 4.2). Both outer segments contain numerous transverse membranous disks (Sjöstrand, 1958; Cohen, 1963), each of which consists of two membranes 6.5–7.0 nm thick enclosing an aqueous intradisk space of 1–3 nm. Thus, total disk thickness is about 14–16 nm, and the disks are separated

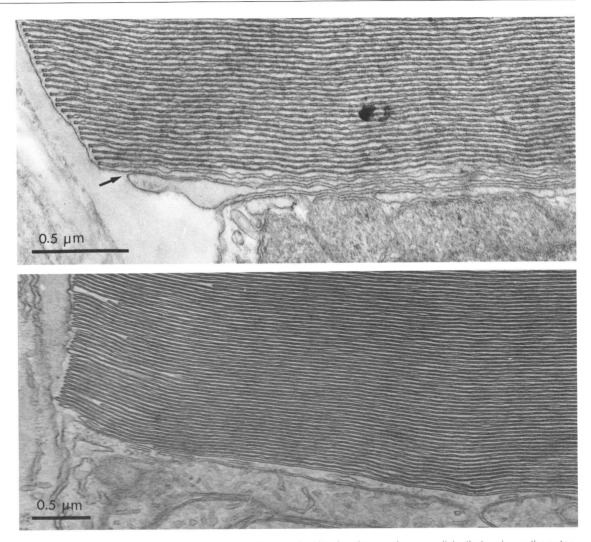

7.3 Electron micrographs showing the membranous disks that make up the outer segments of rods and cones in the mudpuppy. In rods (*top*) most of the disk membranes are separate from the plasma membrane surrounding the outer segment; only a few of the basal disks remain confluent with the plasma membrane (*arrow*). In the cones (*bottom*) the membranous disks remain confluent all along the length of the outer segment.

by a space of about 10–12 nm. In mudpuppy rod outer segments, there are about 1,100 disks; in other outer segments there may be between 500 and 2,000 (Brown et al., 1963; Dowling, 1967).

In the rod outer segment the disks appear to be separate from the plasma membrane; high-magnification electron micrographs (Figure 7.4) show no obvious connection between the disk membranes and the plasma membrane. In the cone outer segment, however, continuity of disks with the plasma membrane is seen along the entire length of the outer segment (Figure 7.3). The fact that the rod disks are not attached to either the plasma membrane or to each other has important consequences for the excitation of the

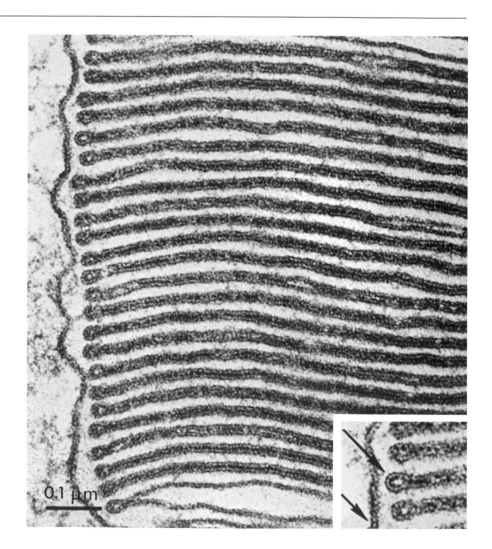

7.4 High-magnification electron micrographs of a portion of a monkey rod outer segment. No connections between the membranous disks and plasma membrane are ordinarily seen. The double-track structure of cell membranes is clearly resolved in these micrographs (*arrows in inset*). From Dowling (1967), reprinted with permission of Harper & Row.

0.1 μm

rod, namely, transfer of information from disk to plasma membrane requires some sort of a messenger (Baylor and Fuortes, 1970).

The visual pigment molecules are membrane bound and are found mainly in the disk membranes (>99 percent), although some visual pigment is present in the plasma membrane (Jan and Revel, 1973). As already noted there are large numbers of visual pigment molecules present in outer segments—on the order of 10^9 molecules per cell. The rod pigment, rhodopsin, has been studied extensively in a number of species and has a molecular weight of about 40,000 (Hubbard, 1953–54). It appears to extend through the entire thickness of the membrane and is oriented such that the amino-terminal end of the protein part of the molecule is always on the inside of the disk and the carboxy-terminal end on the outside. The molecule does not appear to be rigidly anchored in the lipid membrane and it can both rotate and move laterally in the membrane (Cone, 1972; Brown, 1972; Poo and Cone, 1974). Rhodopsin accounts for about 85 percent of the protein in the disks, and the average distance between two rhodopsin molecules is about 20 nm.

The light sensitivity of the visual pigments is due to a chromophore, vitamin A aldehyde (called retinal), which is bound to the visual pigment protein (called opsin) primarily through its terminal aldehyde group (Wald, 1935, 1955) (Figure 7.5). In the eye this complex of protein and chromophore is sensitive to wavelengths of electromagnetic radiation between about 400 and 700 nm; hence this region of the spectrum is our visible light (Plate 8). The major rod pigment rhodopsin peaks at about 500 nm, whereas the cone pigments in primates absorb maximally in the blue (~450 nm), green (~530 nm), and yellow (~565 nm) regions of the spectrum* (Brown and Wald, 1963, 1964; Marks et al., 1964). Two retinals are known, one that derives directly from vitamin A (retinal$_1$) and a second that has a second double bond in the ring part of the vitamin A molecule (Figure 7.5). This second retinal is called dehydroretinal or retinal$_2$. Visual pigments based on dehydroretinal usually absorb maximally toward the red end of the spectrum. The dehydroretinal rod pigment (called porphyopsin) typically peaks at about 525 nm, whereas the dehydroretinal cone pigments typically peak at about 455, 530, and 620 nm (Marks, 1965). Animals that live in fresh water, such as fishes and some amphibians, employ

---

* In primates, the cone pigment that serves red-sensitive vision absorbs maximally in the yellow region of the spectrum (~565 nm) (Plate 8). However, it is commonly referred to as the red-sensitive pigment.

7.5 The structures of vitamin A, retinal (vitamin A aldehyde), dehydroretinal (retinal₂), and the 11-*cis* isomer of retinal. For clarity, only the constituents of the terminal group containing the oxygen atom (O) are shown in each case. For the rest of the molecule, a bend or line indicates a carbon atom and associated hydrogen atoms.

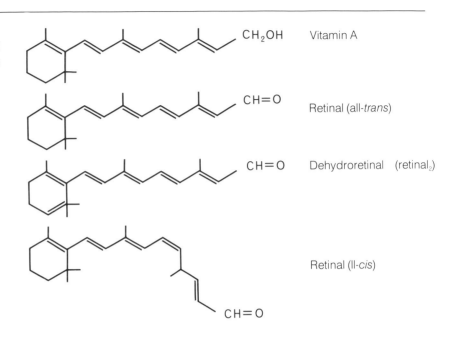

dehydroretinal as their visual pigment chromophore; land animals and marine fishes use retinal₁ (Wald, 1958).

The genes encoding for the opsins of rhodopsin and the red-, green-, and blue-sensitive cone pigments in human have been identified and isolated (Nathans et al., 1986a, b). These experiments showed that different genes specify the opsins (proteins) of the four classes of visual pigment in man and that significant homology exists between the various opsins. The deduced amino acid sequences of the cone opsins showed approximately a 40 percent identity with rod opsin. The red- and green-sensitive pigments whose genes are on the X chromosome, showed a 95 percent identity with each other, but only a 43 percent identity with the blue-sensitive pigment, whose gene is on a different chromosome. Examination of color-blind individuals showed loss or alteration of one or another of the genes. Red-blind individuals (protanopes) were found to have an altered gene for the red-sensitive pigment, whereas green-blind individuals (deuteranopes) had an altered green-sensitive pigment gene or were lacking it altogether. Thus, either color-blind individuals are lacking one or another of the cone pigments or they make an altered pigment (Rushton, 1963a, 1965a).

When a visual pigment molecule absorbs a quantum of light, several molecular transformations occur, first in the chromophore

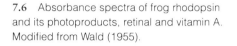

7.6 Absorbance spectra of frog rhodopsin and its photoproducts, retinal and vitamin A. Modified from Wald (1955).

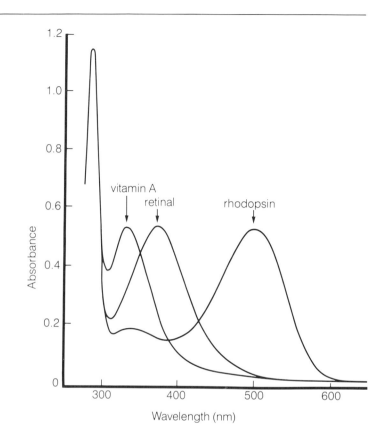

and then in the protein (opsin) part of the molecule (Hubbard and Kropf, 1958). These transformations lead eventually in vertebrates to the separation of the retinal chromophore from opsin. This process is called bleaching because it results in a loss of color of the molecules (Plate 7). Figure 7.6 shows the light absorption properties of a solution of frog rhodopsin before and after bleaching. The native pigment absorbs maximally at about 500 nm; following light exposure, the 500-nm peak is lost and is replaced by a peak at 380 nm, the absorption peak of free retinal. When retinal is converted to vitamin A, which can be done in solution by adding a reducing agent, the retinal peak is lost and is replaced by the vitamin A peak at 328 nm. The absorption observed in the ultraviolet region of the spectrum (approximately 280 nm) is due to opsin itself, and this peak does not change significantly during the bleaching process. (In the intact eye the lens serves as a cutoff filter, limiting short wavelength light (<400 nm) from reaching the retina.

Thus, the ultraviolet absorption by opsin is of no consequence in vision.)

Vitamin A and retinal, because of the alternating single and double bonds in the side chain of the molecules, can exist in a variety of cis–trans configurations, that is, in different shapes (Hubbard and Wald, 1952–53). All visual pigments require one particular cis-isomer, 11-*cis,* for their synthesis (Figure 7.5), and this form of the chromophore combines spontaneously with opsin to form visual pigment (Wald and Brown, 1950; Brown and Wald, 1956). When the visual pigment molecule absorbs a quantum of light, the first transformation is the isomerization of the chromophore from 11-*cis* to all-*trans;* and consequently the chromophore changes shape. The chromophore shape change initiates a series of conformational changes in the opsin, changes that lead ultimately to excitation of the rod cell and the separation of the chromophore from opsin.

Careful study of the absorption changes occurring in visual pigment extracts at different temperatures following a flash of light has enabled investigators to identify at least six intermediates between the absorption of a light quantum by retinal to the release of the chromophore from opsin. There is evidence that many of these intermediates reflect different conformational states of opsin. Only three such intermediates have been detected in intact retinas, but there is no reason to believe that the other intermediates do not occur. The three intermediates detected *in vivo* are relatively long-lived, however, lasting from a fraction of a millisecond to many minutes. Thus, they can be studied and related to physiological events.

Figure 7.7 shows the sequence of rhodopsin-related events that have been observed in the intact retina. After absorbing a quantum, rhodopsin is converted to metarhodopsin I within microseconds at ordinary (room) temperatures. This transition appears to give rise to the R_1 component of the early receptor potential (Cone and Cobbs, 1969). Metarhodopsin I rapidly decays to metarhodopsin II, within a fraction of a millisecond, and this transition gives rise to the R_2 component of the early receptor potential. Metarhodopsin II is much longer lived, with a half-life of several minutes at room temperature. Thereafter, the sequence of events is less clear (Matthews et al., 1963; Cone and Cobbs, 1969; Brin and Ripps, 1977). At least some of metarhodopsin II is converted to a third intermediate, pararhodopsin, but some may break down directly to all-*trans* retinal and opsin. Pararhodopsin is also long-lived and

7.7 Scheme of the sequence of rhodopsin-related events that have been observed in the intact retina. The R_1 wave of the early receptor response (ERP) appears to relate to the transition of rhodopsin to metarhodopsin I; the R_2 wave to the subsequent transition of metarhodopsin I to metarhodopsin II. Metarhodopsin II may break down directly to all-*trans* retinal and opsin, or it may be converted to pararhodopsin. Pararhodopsin may break down to all-*trans* retinal and opsin (question mark), or it may be converted back to metarhodopsin II. All-*trans* retinal may be reversibly converted to all-*trans* vitamin A or be isomerized to 11-*cis* retinal, in which case it may combine spontaneously with opsin to reform rhodopsin. Alternatively, 11-*cis* retinal may be reduced to 11-*cis* vitamin A. Little vitamin A is found in the retina; most of it is transported to the pigment epithelium and stored there (indicated by the dotted line). Absorbance maxima are approximate values, derived mainly from work on rat, frog, and skate retinas.

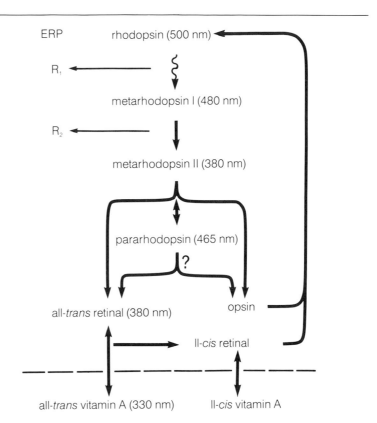

may be observed in a retina for more than 30 min after a bright flash has been delivered to the retina (Frank and Dowling, 1968).

Eventually all of the retinal is released from the opsin, and the retinal may be isomerized back to the 11-*cis* form, in which case it joins spontaneously with opsin to reform rhodopsin, or it may be enzymatically converted to vitamin A for storage purposes. In continuous bright light most of the released retinal is reduced to vitamin A and stored in the pigment epithelium until called upon for regeneration of visual pigments during dark adaptation (Dowling, 1960). Figure 7.8 shows measurements of retinal and vitamin A in both the retina and pigment epithelium during light and dark adaptation and demonstrates the exchange of vitamin A that occurs between retina and pigment epithelium during the process of light and dark adaptation.

Which visual pigment intermediate triggers excitation of the photoreceptor cell? We know that the a-wave begins during the R_2

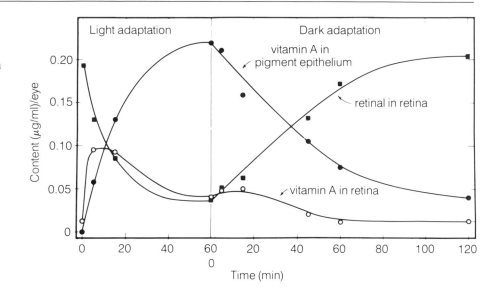

7.8 Distribution of retinal and vitamin A in the rat eye during light and dark adaptation. During light adaptation the retinal content of the retina (*filled squares*) falls as the retinal, liberated by the bleaching of rhodopsin, is reduced to vitamin A. Vitamin A in the retina (*open circles*) rises for a time, but then declines as the bulk of vitamin A moves into the pigment epithelium (*filled circles*). During dark adaptation these processes are reversed. The retinal content of the retina increases as rhodopsin is reformed; and, reciprocally, the vitamin A level in the pigment epithelium declines. The vitamin A content of the retina rises slightly during the onset of dark adaptation but then decreases to the low level found in the dark-adapted retina. From Dowling (1960), reprinted by permission from *Nature* 188:114–118, copyright © 1960, Macmillan Journals Ltd.

component of the early receptor potential. Thus, excitation must be initiated by the time metarhodopsin II appears; and evidence has supported the suggestion that metarhodopsin II serves a key role in excitation (Emeis et al., 1982). The nature of the steps from light-activated rhodopsin to excitation of the visual cell has been clarified significantly in recent years. As discussed earlier, in the dark the membrane surrounding the outer segment is highly permeable to Na^+; light decreases the conductance of the outer segment membrane, thus causing the cell to hyperpolarize. Because most of the rhodopsin is located in the disk membrane and because the conductance change occurs in the plasma membrane, it has long been realized that rhodopsin must release or affect a chemical messenger that controls plasma membrane permeability. There is also an amplification involved between the bleaching of a visual pigment molecule and the reduction in plasma membrane conductance. That is, the bleaching of one rhodopsin molecule leads to the closing of many Na^+ channels (Hagins, 1979). This consideration also indicates the need for an internal messenger system in both rods and cones.

The internal messenger that controls plasma membrane permeability is cyclic GMP. In the dark, cyclic GMP maintains channels in the outer segment membrane in an open configuration (Fesenko et al., 1985; Haynes and Yau, 1985). It does this directly; no phosphorylation or dephosphorylation of the channels appears to be involved. When a photoreceptor cell is illuminated, cyclic GMP levels fall, thereby causing the cyclic GMP-activated channels in the outer segment membrane to close and the cell to hyperpolarize. When cyclic GMP levels increase in photoreceptors, channels open and the cell depolarizes (Lipton et al., 1977; Miller and Nicol, 1979).

The fall in cyclic GMP levels in the light is mediated by the enzyme phosphodiesterase (PDE), which breaks down cyclic GMP to an inactive product, GMP (Miki et al., 1973; Woodruff and Bownds, 1979). PDE, in turn, is activated by a protein, called transducin (or G-protein), which interacts with photoactivated rhodopsin (Fung and Stryer, 1980; Kühn et al., 1981). The interaction of photoactivated rhodopsin (probably metarhodopsin II) with transducin leads to the exchange of a molecule of guanosine diphosphate (GDP) for a molecule of guanosine triphosphate (GTP) on the transducin molecule. When associated with GTP, transducin activates PDE.

Photoactivated rhodopsin dissociates relatively rapidly from transducin and is free to interact with other transducin molecules, until it is inactivated by phosphorylation (Kühn and Dreyer, 1972; Liebman and Pugh, 1980; Kühn et al., 1984). Thus one photoactivated rhodopsin molecule can interact with many transducin molecules ($\sim$500), and amplification of the signal results (Fung et al., 1981). Further amplification occurs in the degradation of cyclic GMP by PDE. One PDE molecule can break down about 2,000 cyclic GMP molecules per second. The cascade of reactions between photon absorption and cyclic GMP inactivation can result in amplification of about 10^6.

The channels in the outer segment membrane controlled by cyclic GMP are highly permeable to Na^+. However, Ca^{2+} also can enter the cell through these pores and appears to play an important regulatory role in the phototransduction process. Ca^{2+}, like transducin, activates PDE, but it inhibits the enzyme, guanylate cyclase (GC), that catalyzes the synthesis of cyclic GMP (reviewed in Stryer, 1986). When the channels in the outer segment plasma membrane close during illumination, Ca^{2+} entry into the cell is shut down and intracellular Ca^{2+} levels fall (Yau and Nakatani, 1985; Gold, 1986). This has two effects: PDE activity is decreased and GC activity is increased. Both of these effects counter the effect of light and serve to increase cyclic GMP levels. During light adaptation of most photoreceptors, membrane potential returns partially toward

7.9 A summary diagram of interactions occurring in the rod outer segment during phototransduction. Light-activated rhodopsin (Rh*) activates transducin (T), which in turn activates the enzyme phosphodiesterase (PDE). These interactions occur in the disk membrane. The activation of PDE leads to the breakdown of cyclic GMP (cGMP) to an inactive product (GMP). Cyclic GMP maintains channels in the outer segment membrane in an open configuration, thereby allowing both Na^+ and Ca^{2+} to enter the cell in the dark. With a fall in cyclic GMP levels in the light, the channels in the outer segment membrane close. The resulting fall of Na^+ levels causes the cell to hyperpolarize. The decrease of Ca^{2+} levels depresses PDE activity and enhances guanylate cyclase (GC) activity, actions that counter the effects of light and increase cyclic GMP levels in the outer segment.

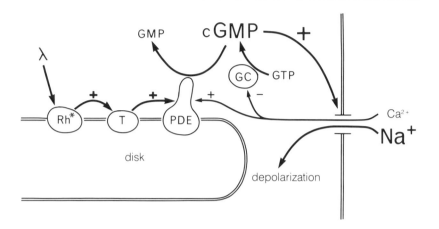

dark levels (see Figure 7.14a), a change that may reflect the increase in cyclic GMP levels induced by lowered intracellular Ca^{2+} during illumination.

Figure 7.9 summarizes the principal interactions that occur in the outer segment of the photoreceptor cell when light is absorbed. In short, photoactive rhodopsin activates transducin, which in turn activates PDE. Cyclic GMP levels fall and consequently channels in the outer segment membrane permeable to Na^+ and Ca^{2+} close. The resulting decrease of intracellular Na^+ levels causes hyperpolarization of the cell, whereas the decrease in Ca^{2+} concentration serves mainly as negative feedback on the system and leads to an increase in cyclic GMP levels. This latter effect may play an important role in light adaptation of photoreceptors.

Photochemical Adaptation

As already noted, the slow recovery of visual threshold following exposure to an intense light appears to be governed by the visual pigments (Figure 7.2). That is, the extent of the slow component of dark adaptation depends on the amount of pigment bleached during light adaptation and the subsequent recovery of threshold appears to relate to, and run in parallel with, the regeneration of the visual pigments. During this phase of the adaptive process, visual sensitivity clearly is governed by photoreceptor mechanisms.

It has long been known that an isolated retina does not regenerate rhodopsin (Kühne, 1878). Why this is so is not understood, but it appears that the outer segments must be in close contact with

7.10   The effects of light adaptation on rhodopsin content (*filled circles*), increment threshold sensitivity (*open circles*), and dark adaptation (*crosses*) determined by measurements made on isolated rat retinas maintained alive. From Weinstein et al. (1967), reprinted by permission from *Nature* 215:134–138, copyright © 1967, Macmillan Journals Ltd.

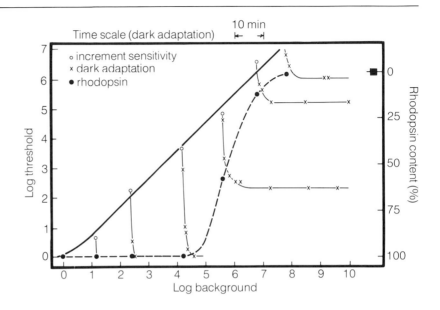

the pigment epithelium and its processes for the isomerization of all-*trans* retinal to the 11-*cis* form to occur. Because it cannot regenerate rhodopsin, the isolated retina is an ideal preparation in which to study the effects of pigment bleaching on visual threshold. Bleaching of rhodopsin in such a preparation should result in a stable and permanent change in the sensitivity of the retina.

Figure 7.10 shows that these expectations are realized (Weinstein et al., 1967). The experiments illustrated are similar to those shown in Figure 7.2, except that here, isolated rat retinas were maintained alive and mounted directly in a recording spectrophotometer. Thus, changes in rhodopsin concentration and b-wave threshold could be monitored virtually simultaneously in the same preparation. As was the case for the intact eye, the rise of increment threshold as a function of adapting intensity was independent of the amount of rhodopsin bleached. Also the rapid (neural) component of dark adaptation was similar to that observed in the intact eye. Thus, when trivial amounts of pigment were bleached by the adapting light (log I_B = <5), b-wave thresholds rapidly returned to dark-adapted levels. When a substantial fraction of rhodopsin was bleached by the adapting light, however, b-wave thresholds fell during dark adaptation to an intermediate level determined by the amount of pigment bleached, and they remained at that level for as long as the experiment lasted. No slow (photochemical) component of dark adaptation like that seen in the intact eye (Figure 7.2) was seen with the isolated rat retina.

From measurements of this sort, it was possible to show that a linear relationship exists between final log b-wave threshold (log sensitivity) and the fraction of pigment in the intact state after bright-light exposure of the isolated retina (Figure 7.11). As already noted a similar relationship between log visual threshold and fraction of bleached pigment has been found to hold during photochemical dark adaptation in rats, humans, skates, and cats (Dowling, 1960; Rushton, 1961; Dowling and Ripps, 1970; Ripps et al., 1981). In all cases the data fit the log-linear expression reasonably well:

$$\log \frac{I_t}{I_0} = k \frac{C_0 - C_t}{C_0},$$

where C_t and I_t are, respectively, the rhodopsin concentration and threshold intensity at time t during dark adaptation, C_0 and I_0 represent these parameters in the fully dark-adapted eye, and k is a constant of proportionality. Where species and rods and cones appear to differ is in the proportionality constant, that is, k is approx-

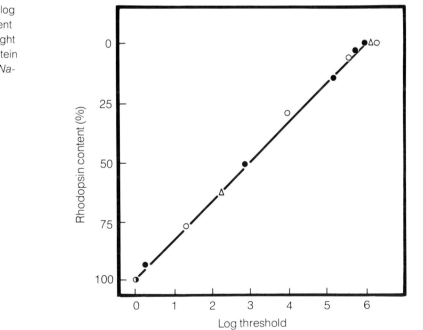

7.11 The linear relationship between final log b-wave sensitivity and percentage of pigment remaining in the rat retina following bright light exposure of the isolated retina. From Weinstein et al. (1967), reprinted by permission from *Nature* 215:134–138, copyright © 1967, Macmillan Journals Ltd.

imately 3 for human cones, but 19 for human rods (Rushton, 1965b). Other proportionality constants are, for rat rods, 5–6; for skate rods, 5; and for cat rods, 6.

In the intact eye visual sensitivity can also be decreased by withholding vitamin A from an animal or a human subject. During the course of the normal breakdown and resynthesis of visual pigment in the eye, small amounts of retinal or vitamin A are lost and must be replenished via the circulation if vision is to be maintained. Indeed, a relationship between nutritional (vitamin A) deficiency and poor dim-light vision (night blindness) has been recognized since the time of the ancient Egyptian medical papyri. If a rat is maintained on a vitamin A-deficient diet, b-wave threshold begins to rise shortly after liver stores of the vitamin are exhausted (Dowling and Wald, 1958, 1960). The rise of threshold occurs simultaneously with rhodopsin depletion, and the relationship observed is a log-linear one. When vitamin A is again introduced in the diet of an animal, log threshold rapidly returns to normal levels as visual pigment is regenerated.

In an isolated retina that has been partially bleached, it is possible to induce rhodopsin regeneration by the direct application of 11-*cis* retinal to the photoreceptors. As expected, this procedure permits the light sensitivity of the retina to recover (Pepperberg et al., 1978). Figure 7.12 shows such an experiment using the all-rod retina of the skate. As already noted, the isolated retina cannot isomerize all-*trans* retinal to the 11-*cis* form; thus, application of all-*trans* retinal to the partially bleached, isolated retina was without effect. However, within a few minutes of the application of 11-*cis* retinal to the partially bleached, isolated retina, substantial recovery of threshold occurred.

In these recovery experiments, thresholds were determined by measuring the aspartate-isolated receptor potential (see Figure 6.11). When receptor thresholds were compared to the amount of rhodopsin present in the retina following partial bleaches, an approximate log-linear relationship was found; and the observed relationship was similar to that observed between visual pigment levels and psychophysical, b-wave, or ganglion cell thresholds during photochemical dark adaptation. Thus, it is clear that bleaching of visual pigment causes a disproportionate rise of visual threshold in receptors as well as in proximal cells. If the only factor limiting threshold was the reduced quantum-catching ability of the receptor as the rhodopsin levels are reduced, the relation between threshold and pigment would follow the dotted line drawn in Figure 7.13. A

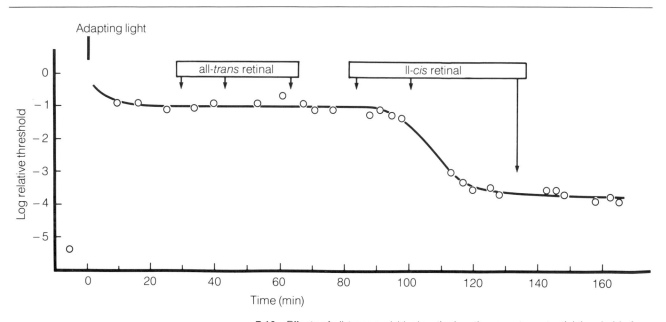

7.12 Effects of all-*trans* and 11-*cis* retinal on the receptor potential threshold of a strongly light-adapted, isolated skate retina. Before time 0, the adapting light bleached about 97 percent of the rhodopsin initially present and raised the threshold of the receptor potential over 4 log units (from log I = −5 to log I = −0.8). At the times indicated by the arrows, all-*trans* or 11-*cis* retinal, suspended in Ringer solution, was applied to the retina. From Pepperberg et al. (1978), reprinted with permission of the Rockefeller University Press.

loss of 50 percent of the visual pigment decreases the quantum-catching ability of the photoreceptors by half (in other words, visual threshold would increase by 0.3 log units), whereas a loss of 90 percent of the visual pigment decreases quantum-catching ability by 90 percent (and would induce a threshold rise of 1 log unit). But, as shown in Figure 7.13, bleaching 50 percent of the visual pigment causes the receptor thresholds to rise about 2.5 log units, whereas bleaching 90 percent of the pigment causes the threshold to rise over 4 log units.

It is of interest to note that the data shown in Figure 7.13 deviate somewhat from an exact log-linear relationship, and these differences do not appear to be due to experimental error. Below about a 50 percent bleach, the experimental points all fall above the log-linear line drawn in Figure 7.13; whereas, with larger bleaches, the data points fall somewhat below the line. Recording intracellularly

7.13   Relationship between rhodopsin content and final log sensitivity of receptor potentials measured extracellularly from the skate retina (*open circles*), intracellularly from toad rods (*filled circles*), and extracellularly from single toad rods with suction electrodes (*filled squares*). All of these data systematically deviate from an exact log–linear relationship (*solid line*). That is, with bleaches below about 50 percent the data points fall above the line, whereas with larger bleaches the data points fall below the line. The dashed line in the figure plots the rise of threshold expected if the only factor limiting threshold were the quantum-catching ability of the receptor as rhodopsin levels are reduced. Data from Pepperberg et al. (1978), open circles; K. N. Leibovic and J. E. Dowling (unpublished), filled circles, and Cornwall et al. (1983), filled squares.

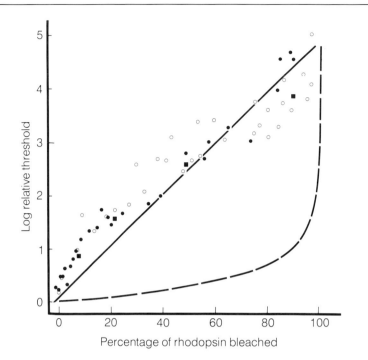

from toad rods, K. N. Leibovic found a similar result, and his unpublished data are shown in Figure 7.13, along with those of Cornwall et al. (1983), who recorded from toad rod outer segments with suction electrodes. Whether there are slight differences in the relation of visual pigment levels and light thresholds in photoreceptors and more proximal cells is not clear, and how precisely the log-linear relation will turn out to fit all of these data remains to be seen. Nevertheless, the data are unequivocal in showing that bleached visual pigment profoundly alters receptor sensitivity and, consequently, the sensitivity of proximal neurons.

Can any other component of visual adaptation be related directly to photochemical events? Some years ago it was proposed that the fast initial phase of the dark adaptation process reflects the decay of the rhodopsin intermediates in the outer segments of the photoreceptors (Donner and Reuter, 1968). In both the rat and the skate, experiments have been carried out comparing the time course of intermediate decay with b-wave or receptor potential thresholds (Frank and Dowling, 1968; Brin and Ripps, 1977). No relation was observed between rhodopsin photoproduct decay and light sensitivity. Thus, the fast components of both dark adaptation and light adaptation appear to be governed by factors other than the photochemistry.

Photoreceptor Light Adaptation

As discussed in Chapter 4, vertebrate photoreceptors have an intensity–response relationship that spans slightly more than 3 log units. This means, of course, that with light flashes greater in intensity than 3.5 log units above threshold, the photoreceptor response amplitude is maximal, and no larger responses can be elicited from the cell regardless of stimulating intensity. On the other hand, there is abundant evidence that most photoreceptors continue to respond to photic stimuli when the receptor is continuously illuminated with lights considerably brighter than 3.5 log units above their dark-adapted threshold (Boynton and Whitten, 1970; Dowling and Ripps, 1970; Kleinschmidt, 1973; Baylor and Hodgkin, 1974). How do the photoreceptors manage to respond when illuminated with lights far above their amplitude saturation level?

There are two mechanisms that appear to operate during photoreceptor light adaptation and allow the cell to continue to respond in bright light (Dowling and Ripps, 1970; Kleinschmidt, 1973; Baylor and Hodgkin, 1974; Kleinschmidt and Dowling, 1975; Fain, 1976). First, when receptors are illuminated with continuous light, the initial hyperpolarizing response is not maintained. With time the membrane potential gradually returns toward dark levels, so that with moderate and bright background lights, this adaptive mechanism brings the receptor below its saturation level. If the saturation level were maintained, on the other hand, the receptor cell would be unresponsive to increment flashes of any intensity. In background light the membrane potential never recedes all the way back to dark-adapted resting potentials; rather it stabilizes at a plateau of hyperpolarization whose magnitude is graded with background intensity. (As noted earlier, this adaptive mechanism may relate to the action of Ca^{2+} in the phototransduction process.)

This phenomenon is illustrated in Figure 7.14 with data from gecko rods. Figure 7.14a shows the recovery of membrane potential following the application of a sustained moderately bright light ($\log I = -4.2$) and the increment responses elicited with fixed-intensity test flashes for the first 10–15 sec after the light was turned on. The two records, from normal and aspartate-treated eyecups, show virtually identical results. Following the onset of illumination, membrane potential recovered by more than 50 per-

7.14 Light adaptation of gecko photo-receptors.

a Intracellular responses from photoreceptors in control and aspartate-treated retinas following the onset (and maintenance) of an adapting light. The record on the left was made from a normal eyecup; the record on the right was made from an aspartate-treated eyecup in which the receptors were isolated from the influences of the horizontal cells. Log background = −4.2; test flash log intensity = −3.5.

b Graph illustrating the V–log *I* curves of the peak and plateau responses of dark-adapted gecko photoreceptors (*thick lines*) and V–log *I* curves of the peak responses determined in the presence of two different adapting lights (log *I* = −4.2 and −2.2) (*thin lines*).

c V–log *I* curves for the peak responses of gecko photoreceptors plotted on log–log coordinates. Measurements were made on photoreceptors in the dark-adapted state (DA) and in the presence of two background intensities (log *I* = −4.2 and −2.2). Modified from Kleinschmidt and Dowling (1975), with permission of the Rockefeller University Press.

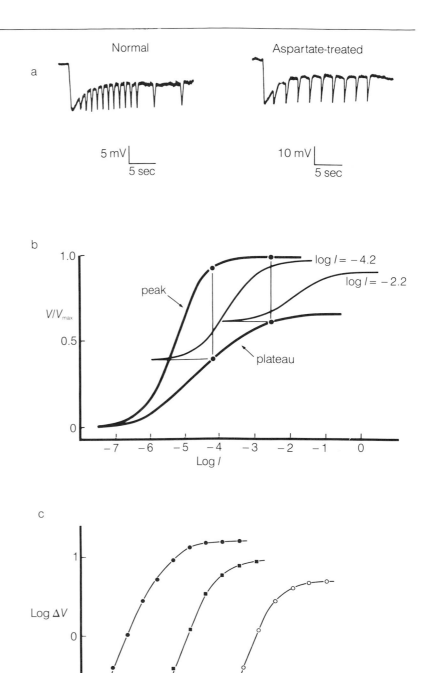

cent from the initial level of hyperpolarization in about 5 sec. As membrane potential returned toward dark levels, the increment response amplitudes to the test flash increased correspondingly. That the same results were observed in the normal and aspartate-treated eyes means that this adaptive mechanism is intrinsic to the receptors and is not caused by, or influenced significantly by, horizontal cell feedback to the receptors.

In Figure 7.14b, the initial (peak) hyperpolarizing response of the gecko rods to a wide range of background lights (log I), and the steady-state (plateau) hyperpolarizations, are represented by thick lines. At all intensities the plateau potential is considerably below the peak potential, and it levels off at about 65 percent of the saturated response amplitude level.

The second adaptive mechanism in the receptor shifts the photoreceptor intensity–response curves along the intensity axis, thus extending the range over which the receptor responds (Figure 7.14b and c). In Figure 7.14b the increment V–log I curves (thin lines) are plotted relative to the plateau level upon which the increment responses are superimposed. The V–log I curves extend over the same range as the dark-adapted V–log I curve (~ 3 log units), and they saturate at about the same voltage level as the dark-adapted curve. However, they are compressed because they arise from the partially hyperpolarized, plateau level. In Figure 7.14c the three V–log I curves are plotted on log–log coordinates. The shift of the curves on the intensity axis is clear, as is the fact that the curves do not change shape; that is, at all background intensities the receptor potentials have an initial slope of unity and are thus linearly related to light intensity.

Figure 7.15 shows the increment threshold function for gecko photoreceptors, derived by determining the intensity of light necessary to evoke a 0.5-mV incremental response at various backgrounds. As can be readily seen, the photoreceptors adapt to background lights over a wide range of intensities in conformity with the Weber-Fechner relationship; that is, threshold varies in direct proportion to background intensity. Thus, receptors adapt to background light very much as do psychophysical thresholds, b-waves, and ganglion cell discharges (Rushton, 1965b; Dowling and Ripps, 1970). The data shown in Figure 7.15 were derived from both normal eyecups and eyecups treated with aspartate, so the conclusion can be drawn that both components of photoreceptor adaptation (that is, membrane potential recovery and lateral shift of the V–log

7.15   Incremental thresholds of normal and aspartate-treated gecko photoreceptors over a range of background intensities. Incremental thresholds were determined by measuring the light intensity needed to evoke a 0.5-mV potential intracellularly from a cell. From Kleinschmidt and Dowling (1975), with permission of the Rockefeller University Press.

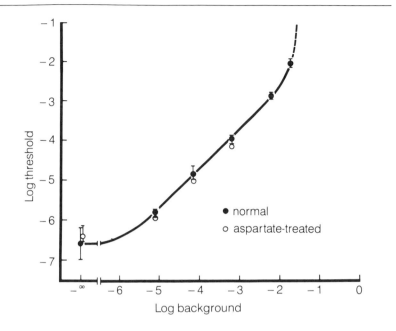

I curves on the intensity axis) are intrinsic to the cell and not a result of feedback influences.

With bright backgrounds ($>\log I = -2$), thresholds of the gecko photoreceptors deviate from the Weber-Fechner relationship and rise rapidly to saturation with small additional increases in background luminance. Saturation occurs in the gecko receptor even though the steady-state plateau level is well below the maximum amplitude level of the cell. Why no further responses can be elicited from the cells under these conditions is not entirely clear. Rod saturation occurs well before a substantial fraction of visual pigment has been bleached, so loss of quantum-catching ability by the photoreceptor is not a factor. Rods may saturate even though rod membrane potential is well below maximum amplitude level because two mechanisms can contribute to membrane potential recovery in the light. Part of the membrane potential recovery results from reopening of the light-sensitive Na^+ channels in the outer segment membrane. But opening of voltage-sensitive channels in the inner segment membrane can also reduce membrane potential levels of the receptor in the light (see Normann and Perlman, 1979a). Saturation may occur in the rods because all of the light-sensitive channels are closed.

The saturation of rod response in bright background illumination is well known and has been shown to occur in the rods of humans (Aguilar and Stiles, 1954), rats (Green, 1973), mudpuppies (Normann and Werblin, 1974), and toads (Fain, 1976). The only known exceptions are the rods of the skate, which continue to give incremental responses at 8 log units above absolute threshold (Dowling and Ripps, 1972). The skate, however, is unique in that it has an all-rod retina. If the skate is to continue to see in bright light, its receptors cannot saturate. All other species have cones, and above rod saturation cones mediate visual responsiveness. Cones, unlike rods, do not appear to saturate; they continue to give responses regardless of background intensity (Boynton and Whitten, 1970; Normann and Werblin, 1974; Baylor and Hodgkin, 1974; but see Naka et al., 1979).

When evaluated on the basis of adaptation to background lights, saturating rods appear to be of two types. Some rods (in the gecko and the toad) show a significant amount of adaptation; that is, they shift their voltage–intensity function by about 4 log units before they saturate, and they obey the Weber-Fechner relationship over a range of at least 3 log units. Other rods (in the mudpuppy and the monkey) adapt very little. That is, they do not significantly shift their voltage–intensity curves in response to backgrounds, and their responses are thus limited to the narrow dynamic range of the dark-adapted receptor. Incremental thresholds rise only 2–3 log units before the receptor saturates, and they never really conform to the Weber-Fechner relationship.

The two mechanisms that underlie light adaptation of the receptors—namely, the partial recovery of membrane potential in the light and the lateral shift of the voltage–intensity curve—appear to be relatively independent of one another. This conclusion derives from experiments on the skate retina, where the membrane potential recovery in the light is very slow, lasting up to 30 min or more following the onset of bright adapting lights (Dowling and Ripps, 1971, 1972). Thus, complete voltage–intensity curves can be plotted at will during the recovery of response amplitude in the light. Figure 7.16 shows the results of a typical experiment. The first voltage–intensity curve was determined 5 min after the onset of the adapting light; it was shifted to the right on the intensity axis but was very restricted in terms of amplitude range. With time, however, maximum receptor response amplitudes grew very significantly, and by 25 min, they were about 60 percent of the amplitude of the maximum dark-adapted potential. The voltage–intensity

7.16 Recovery of extracellularly recorded receptor potential amplitude over time following the onset of an adapting light in the skate. Modified from Dowling and Ripps (1972), with permission of the Rockefeller University Press.

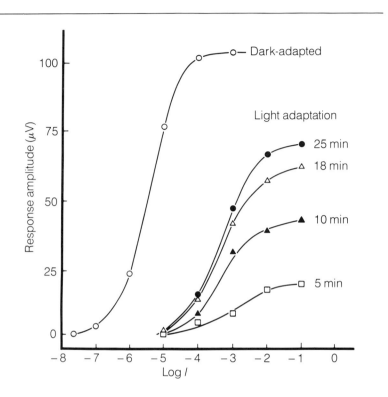

curve, on the other hand, was fully shifted by the time the first voltage–intensity curve was determined, and it remained fixed in position even though the voltage–intensity curves grew.

The increase of receptor response amplitude with time in the light clearly reflects a recovery of membrane potential (Figure 7.14a) (Dowling and Ripps, 1971; Kleinschmidt and Dowling, 1975). The shift of the voltage–intensity curve, on the other hand, appears to be independent of membrane potential (Figure 7.16) (but see Itzhaki and Perlman, 1987). Thus, it may reflect a mechanism close to the initial phototransduction events; that is, it may be related to the biochemical interactions that occur on or near the disk membrane. Evidence in favor of this notion has come from experiments indicating that light adaptation of the rod also requires an internal messenger system (Bastian and Fain, 1979). A weak background light that over a period of 10 sec bleaches just 40 rhodopsin molecules reduces the sensitivity of toad rods (that is, response amplitude to dim flashes) by about 50 percent. Because there are about 2,000 disks in a toad rod, fewer than 2 percent of the disks can have a bleached rhodopsin molecule in such an experiment. Thus, bleached rhodopsin molecules in one disk membrane must be able to depress the effects of excited rhodopsin molecules in other disk membranes. Careful measurements of the spread of desensitization along toad rods in the light indicate that the adapting message extends 5–10 μm along the rod from the site of bleached pigment (Lamb et al., 1981).

It is interesting to note, finally, that during much of slow photochemical dark adaptation, membrane potential of the receptor is close to or at dark levels (Kleinschmidt and Dowling, 1975). So the decreased light sensitivity that results from visual pigment bleaching also appears to be governed by a mechanism remote from the plasma membrane. But this desensitization does not appear to spread from the site of pigment bleaching; in other words, no internal messenger is involved (Cornwall et al., 1983). Thus, bleaching desensitization in the rod appears to be different from background light desensitization.

Network Mechanisms

It is clear that photoreceptors show adaptive properties similar to those exhibited in proximal neurons and that, at least during slow photochemical dark adaptation, sensitivity all along the visual system is governed by the photoreceptors. It is well established, however, that under certain conditions nonreceptoral or network mechanisms of adaptation are also capable of altering the light sensitivity of retinal elements and governing the visual adaptation process (see Rushton, 1963b, 1965b). A particularly striking and clear example of this was found in the skate retina and is shown in Figure 7.17 (Green et al., 1975). In these experiments the retina was illuminated with lights so dim that only a fraction of the rods were absorbing a quantum per second. Under such conditions, neither the receptors nor the horizontal cells showed any evidence of adaptation (that is, voltage–intensity curves determined in the dark and in the presence of these weak background fields were identical). On the other hand, the sensitivities of the PNR and b-wave were significantly altered; the voltage–intensity curves recorded in the presence of the dim background fields were shifted about 1 log unit to the right on the intensity axis and the maximum voltage generated was reduced. Because the receptor responses were unchanged by the weak ambient illumination, the adaptive changes occurring in the proximal neurons under these conditions cannot be attributed to receptoral mechanisms.

7.17 Voltage–intensity curves for extracellularly recorded receptor potential, horizontal cell response, b-wave, and proximal negative response (PNR), measured from the skate eyecup preparation in the dark and in the presence of a dim adapting field. Modified from Green et al. (1975) and Dowling and Ripps (1977), both with permission of the Rockefeller University Press.

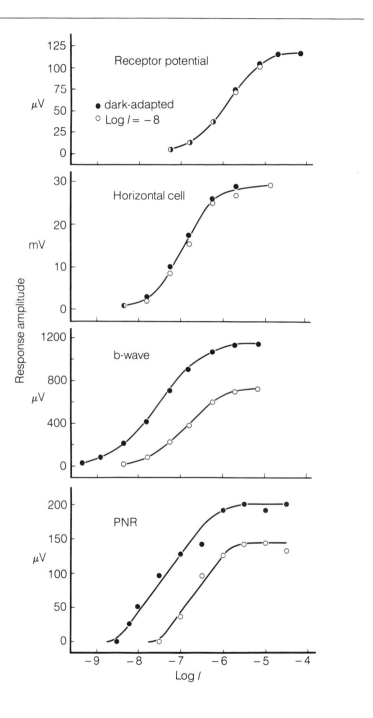

Skate ganglion cells also showed decreased sensitivity when the retina was illuminated with dim lights; and the loss of sensitivity, as a function of background intensity, was found to be the same for all three of the proximal retinal responses measured. The skate horizontal cells, on the other hand, showed adaptive properties very similar to those of the photoreceptors and only began to show decreased sensitivity when the retina was illuminated with background lights 2 log units brighter than those that raised the threshold of the proximal responses. With bright background lights, however, all of the retinal responses exhibited similar threshold changes, and they all followed the Weber-Fechner relationship.

Figure 7.18 shows these relationships. The absolute thresholds of the receptor potential and b-wave are plotted relative to one another; for example, the b-wave absolute threshold is about 2 log units more sensitive to light than is the receptor potential (see also Figure 7.17). Threshold changes due to background illumination of the horizontal cell are plotted relative to receptor potential thresholds, whereas threshold changes of the PNR and ganglion cell discharge are plotted relative to b-wave thresholds. The figure shows that the light sensitivities of the b-wave, PNR, and ganglion cell discharge were all equally and significantly altered by dim background illumination ($<\log I = -6$) that had no or little effect

7.18 Incremental threshold functions for the b-wave, PNR, a ganglion cell, receptor potential, and a horizontal cell, all from the skate retina. The arrow indicates the background intensity at which 1 quantum per second is absorbed per rod, assuming that one-third of the quanta incident on the retina were absorbed. Data from Green et al. (1975) and Dowling and Ripps (1977), both with permission of the Rockefeller University Press.

on the sensitivity of the receptor potential or horizontal cell response. As background illumination was raised, however, the thresholds of the distal and proximal responses converged; and, with backgrounds above $\log I = -6$, similar threshold changes were observed in all of the responses as a function of adapting luminance.

In addition to demonstrating the existence of a nonreceptoral, network adaptive mechanism operating when the retina is exposed to very weak adapting lights, two other important conclusions can be drawn from the data shown in Figure 7.18. First, it seems likely that the receptors govern the light adaptation process for the entire retina under bright background conditions, because with bright backgrounds all of the retinal responses showed similar changes in light sensitivity. Second, the mechanism underlying the network adaptive process must lie between the generation of the horizontal cell response and the generation of the b-wave, because the proximal neurons are affected by weak backgrounds that do not affect the distal responses.

In other situations, it is possible to show that network adaptive mechanisms govern the sensitivity of the proximal retinal responses. For example, Figure 7.19 shows that the early (neural) phase of dark adaptation appears to be mediated in the skate by a nonreceptoral mechanism. Data presented in Figure 7.19a demonstrate that following a weak adapting exposure that did not bleach a significant fraction of the rhodopsin, thresholds for the b-wave and ganglion cell discharge recovered in concert and their rates of recovery were significantly slower than those for either the receptor potential or the horizontal cell. Data presented in Figure 17.9b indicate that following a stronger adapting exposure, which bleached about 50 percent of the rhodopsin, the initial recovery of b-wave threshold was also slower than that of the receptor potential. But after about 10 min the thresholds of the b-wave and receptor potential converged, and they recovered in unison over the last 30 min of dark adaptation, in parallel with the regeneration of rhodopsin.

A summary of the findings in the skate are presented in Figure 7.20. They suggest what the contributions of receptoral and network mechanisms might be to the process of visual adaptation determined at the level of the ganglion cells or psychophysically. For example, threshold changes that result from light adaptation to dim backgrounds are clearly due to network mechanisms, because the receptors exhibit little or no change in sensitivity in these circumstances. With bright backgrounds, receptoral thresholds as well as

7.19 a Dark adaptation of skate receptors, horizontal cells, the b-wave, and a ganglion cell following a light adaptation that did not bleach a significant fraction of the visual pigment in the retina. From Green et al. (1975).

b Dark adaptation of skate receptors and b-wave following a light adaptation that bleached about 50 percent of the rhodopsin in the eye. Modified from Green et al. (1975), with permission of the Rockefeller University Press.

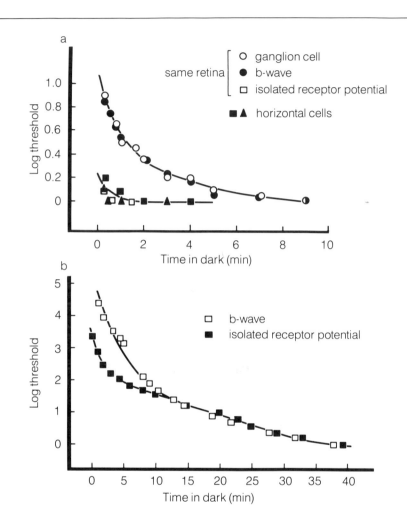

those of more proximal elements increase as a linear function of adapting intensity, a result suggesting that receptors govern the light adapting process under such adapting conditions. During dark adaptation the situation is reversed. The initial threshold recovery reflects a network mechanism (that is, it is the rate-limiting process), whereas the subsequent, very slow recovery of threshold occurring in parallel with rhodopsin regeneration reflects receptoral mechanisms. That is, during photochemical dark adaptation the receptors regulate the sensitivity of all the retinal elements.

Do similar network adaptive mechanisms exist in other retinas? In the rat Green and Powers (1982) have shown that dim back-

7.20   Scheme showing the contribution of receptoral and network adaptive mechanisms to the process of visual adaptation in the all-rod skate retina. *Left*, Incremental threshold function; *right*, a dark adaptation curve made from an eye preexposed to an adapting field that bleaches a substantial fraction of the rhodopsin. The initial part of both the incremental threshold function and dark adaptation curve appears to be governed by network mechanisms; the latter portion of each function appears to be governed by the photoreceptors. From Dowling (1977), reprinted with permission of MIT Press.

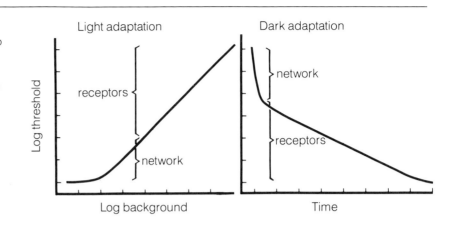

ground lights that do not affect receptor sensitivity decrease b-wave and ganglion cell sensitivity equally and significantly. In humans dim background lights affect rod b-wave light sensitivity more than rod a-wave sensitivity (Fulton and Rushton, 1978), and a similar result has been found in the carp (Witkovsky et al., 1975). In both the carp and the ground squirrel, however, there is a parallel increase in cone b-wave and receptor potential thresholds with increasing intensity, findings suggesting that in the photopic range all of the rise in b-wave threshold affected by background illumination can be accounted for by cone desensitization (Witkovsky et al., 1975; Green and Dowling, 1975). As yet there is no information concerning the locus of the mechanisms underlying the initial, nonphotochemical component of dark adaptation in any species other than the skate.

The kinetics of the network adaptive mechanism has been studied to some extent in the skate retina (Green et al., 1975). Both its onset and recovery (Figure 7.19) were found to be surprisingly slow. Although in each case some changes in sensitivity were observed within seconds, or perhaps even within a fraction of a second, final thresholds were often not achieved for minutes following the onset or offset of an adapting light. Synaptic processes usually do not have such long time constants, a fact suggesting that nonsynaptic events could be responsible for these effects. Illumination could cause a desensitizing agent to accumulate in the retinal network, a change that could affect the sensitivity of one or a number of retinal elements.

In this regard it has been observed that raised extracellular levels of K⁺ will affect b-wave and PNR threshold but not receptor or horizontal cell thresholds (Dowling and Ripps, 1976a; Dowling, 1977). When extracellular K⁺ levels were raised in the skate eyecup and incremental thresholds for the b-wave and receptor potential determined over a wide range of background intensities, a revealing result was obtained (Figure 7.21). First, over an extracellular K⁺ concentration range of 4 to 45 mM, the incremental threshold curve for the receptors was unchanged. However, increased K⁺ levels of just a few millimoles raised the dark-adapted threshold for the b-wave and selectively altered the initial portion of the b-wave incremental threshold function ascribed to the network mechanisms. In other words, as the dark-adapted threshold of the b-wave was raised by increased levels of extracellular K⁺, b-wave sensitivity was less and less affected by weak background lights. Indeed, with Ringer solution containing 18 mM K⁺, the absolute threshold of the b-wave was raised close to that of the receptor potential, and it appeared as if the network adaptive mechanisms were abolished. Thus, under these conditions, the b-wave light-adapted exactly like the receptors, and alterations in b-wave sensitivity at all back-

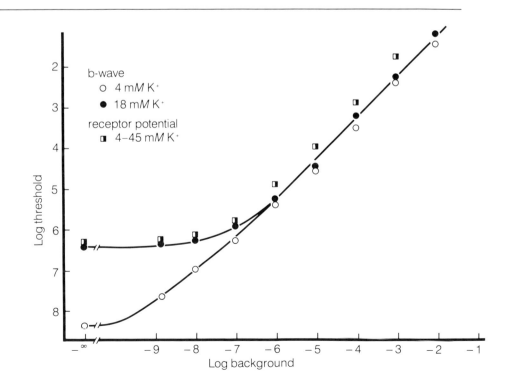

7.21 Incremental threshold functions for the b-wave and receptor potential of the skate, determined in Ringer solution containing various concentrations of K⁺. From Dowling (1977), reprinted with permission of MIT Press.

ground levels could be accounted for by changes in receptoral sensitivity.

These experiments thus provide further evidence that the changes in retinal sensitivity induced by dim lights in the normal eye are due to the network mechanisms whereas alterations in retinal sensitivity induced by bright illumination reflect receptoral changes. They also show that altered extracellular K^+ levels mimic to a considerable degree the effects on retinal sensitivity ascribed to network adaptation.

Synaptic Adaptive Mechanisms

So far the effects of diffuse background lights on the responses of the various retinal elements have been discussed. However, the light sensitivity of bipolar and ganglion cells (and presumably the other retinal neurons) can be further modified by specific forms of background illumination presented in the surrounds of their receptive fields. In Chapter 4, for example, it was shown that by applying surround illumination to the bipolar cell receptive field the intensity–response curve for the center response was shifted to the right on the intensity axis (Figure 4.10); that is, the center response became less sensitive as a result of surround illumination. The extent of the shift was also shown to be intensity dependent; the brighter the surround illumination, the greater its densensitizing effects.

A similar type of desensitizing phenomenon can also be seen to operate at the level of the inner plexiform layer when stimuli specific for activating the transient amacrine cells are presented to the retina and the responses of on–off ganglion cells recorded (Werblin and Copenhagen, 1974). Figure 7.22 shows the results of such an experiment in the mudpuppy retina. An on–off ganglion cell was recorded intracellularly and graded responses evoked with a central spot of light. A windmill-like stimulus was applied to the surround region of the cell's receptive field. When stationary, the stimulus had no effect on the center response. When the windmill was spinning, however, the response–intensity curve was shifted to the right on the intensity axis and the maximum response of the center was decreased. Although the change in sensitivity induced by the spinning windmill was relatively small (~ 0.15 log units), it was clear and distinct.

Changes in on–off ganglion sensitivity were also examined as a function of time following the presentation of an annular field to the retina. The intensity of the background was set so that final increment threshold would be about 2 log units above initial

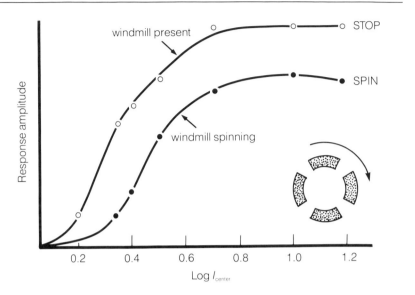

7.22 Effects of a spinning windmill on the amplitude of the graded response of an on–off ganglion cell in the mudpuppy. The intracellular responses were elicited by a central flash of light with the windmill first stationary and then spinning. From Werblin and Copenhagen (1974), reprinted with permission of the Rockefeller University Press.

threshold. Following the onset of the surround illumination, thresholds rose rapidly; and at 250 msec after field onset, they were unmeasurable (that is, they were at least 4 log units above initial thresholds). A rapid recovery of sensitivity occurred thereafter, and final thresholds were achieved in about 1 sec. The time course of the effect, with some delay, is similar to the time course of the response of the transient amacrine cells in the mudpuppy (Figure 4.1).

Not all ganglion cells demonstrate these effects. The sustained on-center ganglion cells, which in the mudpuppy appear to receive little or no input from the transient amacrine cells, are not specifically affected by a spinning windmill. And, as would be expected, the bipolar cells, which appear to provide the major part of the input to the sustained on-center ganglion cells (Chapter 4), are also not affected by a spinning windmill.

In summary, adaptive processes exist throughout the retina, from the receptors to the ganglion cells. The light sensitivity of the various retinal elements at any particular point in time depends not only on background luminance but also on the distribution of light within the receptive field, the dynamics of the stimulus (that is, whether the adapting field is moving or not), and the immediate past history of retinal illumination. Some of the adaptive processes are located within the outer segments of the photoreceptors and appear to be closely associated with the initial phototransduction process whereas others are clearly mediated synaptically. Finally, there are adaptive processes that are nonreceptoral in nature and that may not be mediated synaptically. The ultimate light sensitivity of the ganglion cells, the output neurons of the retina, reflects a variety of mechanisms occurring at various levels within the retina. Furthermore, different types of ganglion cells are likely to demonstrate somewhat different levels of adaptation at any one time depending on the nature of the adapting stimuli.

EIGHT

Retinal and Brain Mechanisms

THERE are many reasons for studying the retina. The first stages of visual processing are carried out there, and, if we are to understand vision and visual perception, knowledge of retinal mechanisms is requisite. The retina also is subject to disease and degenerative processes that lead to blindness, and if we are to aid these problems, understanding of retinal processes is crucial. In this book, I have emphasized that the retina is an approachable part of the brain that can provide significant insights into central nervous system mechanisms in higher animals. In the preceding chapters, much of what we presently know about retinal structure and function has been described, and along the way general models of how the retina works have been formulated. It is appropriate now to ask what these studies of the retina tell us about brain mechanisms and function.

Progress in retinal studies has been remarkable over the past few decades, and the retina may be the best-understood part of the central nervous system in vertebrates. We have a reasonable picture of the pathways of information flow through the retina's two plexiform layers, and we know at least in general terms how each major retinal neuron responds when the retina is presented with light stimuli. Neurotransmitters at key synapses in the retina have been identified, at least provisionally; and we have seen how both neurons and glial cells contribute to potentials generated across the retina. Finally, we have learned that light sensitivity of retinal elements can be modified by mechanisms operating throughout the retina, from receptors to ganglion cells. In this chapter I shall review some of the studies that have had special significance for our understanding of brain mechanisms.

Local-Circuit Neurons

An early and important discovery deriving from analysis of the retina's synaptic organization was that amacrine cell processes are both pre- and postsynaptic (Chapter 3). Not only did these observations immediately suggest how axonless cells in the brain could function (their processes have the characteristics of both dendrites and axons), but they also implied that interactions between amacrine cell processes could occur locally and not involve the entire neuron. Observations in single micrographs of reciprocal and serial synaptic arrangements, along restricted portions of amacrine cell processes, strengthened the notion that synaptic interactions could occur locally in these cells (see Figure 3.15).

At virtually the same time these observations on the retina were being described (Dowling and Boycott, 1966), similar findings were made in the olfactory bulb, another part of the brain where Cajal had long ago pointed out numerous axonless cells (Rall et al., 1966). In the olfactory bulb, axonless granule cells make reciprocal synaptic junctions with dendrites of mitral cells, the output neurons of the olfactory bulb. Synapses made by the mitral cell dendrites onto granule cell processes were observed to have round synaptic vesicles, a finding suggesting that these are excitatory junctions. However, the reciprocal synapses from granule cell processes back onto mitral cell dendrites were characterized by flattened synaptic vesicles, a finding suggesting inhibitory junctions. Physiological experiments (Rall and Shepherd, 1968) subsequently provided evidence that in the olfactory bulb the reciprocal junctions between the mitral and granule cells represent a local inhibitory feedback circuit; when a mitral cell dendrite is depolarized, excitatory synapses onto granule cell processes are activated and the granule cell subsequently depolarized. This depolarization activates the reciprocal inhibitory synapses back to the mitral cell dendrite, and the mitral cell dendrites are inhibited. Furthermore, passive spread of current in the granule cell processes can activate other inhibitory synapses and thereby inhibit additional mitral cells. The spread of inhibition depends, of course, on how strongly the granule cell dendrites are initially depolarized.

Following these initial observations on the axonless cells in the retina and olfactory bulb, investigators found that other cells— particularly those with short axons present in several parts of the vertebrate brain, including the retina (that is, horizontal cells) (Dowling and Werblin, 1969), olfactory bulb (Pinching and Powell, 1971), several thalamic nuclei (Ralston, 1971, 1979), and the motor cortex (Sloper, 1971)—form similar kinds of local synaptic arrangements. Most often these involve dendrites of cells, but instances of complex synaptic arrangements in axonal processes and terminals are also known. These findings renewed interest in distinguishing two basic types of neurons in the vertebrate brain: long-axon neurons, or Golgi type I neurons, which carry information from one part of the brain to another; and short-axon neurons, or Golgi type II neurons, whose processes are confined to a restricted area of the brain (that is, one nucleus or a single plexiform layer) and mediate local interactions between neurons (Rakič, 1975). This distinction was first suggested by Cajal nearly 100 years ago, and he pointed out that short-axon neurons appear to play an important role in mediating complex brain interactions. He stated that "the functional superiority of the human brain is intimately linked up with the prodigious abundance and unaccustomed wealth of forms of the so-called neurons with short axons" (see Ramón y Cajal, 1964, p. 480). In substantiation of this notion, Cajal noted that the ratio of the number of short-axon neurons to the number of long-axon cells was considerably larger in the human brain than in the brain of the mouse or other nonprimate mammals. More recent studies have confirmed Cajal's early observations (Mitra, 1955). The retinal observations that those species with more complex ganglion cell receptive field organizations have relatively more amacrine cells, with more complex amacrine cell synaptic arrangements, fits well with this general notion. Indeed, we have gained some understanding of how amacrine cells in the retina underlie complex receptive field properties such as detection of motion, direction, and orientation.

We are still unclear about the details of how interactions between the short-axon and axonless Golgi type II cells in the retina give rise to complex receptive field properties. In particular, because our electrophysiological recordings are almost exclusively from cell perikarya, and thus reflect activity of the whole cell, we have little notion yet of how local synaptic interactions within the dendritic tree of a short-axon or axonless cell are involved in information processing. Some insights in this regard have been provided by study of the cholinergic amacrine cells in the rabbit retina, which provide excitatory input to the directionally selective ganglion cells (see Chapters 4 and 5 and Masland et al., 1984b). These amacrine cells are relatively large and have dendritic fields that are several hundred micrometers in diameter (Plate 2) and overlap extensively in the inner plexiform layer. Yet, a directionally selective ganglion cell in the rabbit can resolve movements of stimuli much smaller than the size of a single cholinergic amacrine cell. How can it be that a ganglion cell can detect movement of a stimulus of dimensions much smaller than those of the input neuron? An obvious explanation is that in response to a small stimulus acetylcholine is released locally from a restricted number of synaptic sites rather than from all of the cell's synapses, which would be the case in response to a large stimulus (Miller, 1979; Masland et al., 1984b). This idea is supported by the finding that the output synapses of the cholinergic amacrine cells are confined to the distal third of the cell's processes (Famiglietti, 1983), an arrangement that helps to isolate the output of individual processes and promote local release.

Two other points concerning Golgi type II cell function come mainly from studies of the retina and particularly those of the retinal amacrine cells. First, although Golgi type II cells in the brain are usually inhibitory (Rakič, 1975), the cholinergic amacrine cells just described are clearly excitatory and amacrine cells containing substance P have excitatory actions also (Glickman et al., 1982). Thus Golgi type II cells can be excitatory as well as inhibitory. Second, Golgi type II cells can be interposed between Golgi type I cells, an arrangement that allows the input to a Golgi type I cell to emanate entirely from type II cells. This arrangement was clearly demonstrated in the ground squirrel retina, where certain ganglion cells receive input only from amacrine cells (West and Dowling, 1972). In such instances, all of the visual information conveyed to the ganglion cell passes through a Golgi type II cell network, and one presumes that both excitatory and inhibitory cells are involved.

Graded-Potential Neurons

Implied in the thinking about local-circuit neurons and local interactions is the notion that graded synaptic potentials can initiate synaptic activity and mediate information processing within the brain. This idea suggests that neurons could function entirely without action potentials. Indeed, as described in Chapter 4, most of the intrinsic neurons in the retina operate only with graded potentials. Even transient amacrine cells, which do produce action potentials, usually fire only a few in response to a strong stimulus, and it seems that graded potentials generated in the cell are more important for retinal function. For example, Murakami and Shigematsu (1970) applied tetrodotoxin, a potent action potential-blocking substance, to the retina and subsequently recorded light-evoked potentials in ganglion cells, thereby showing that slow potential activity can mediate synaptic interactions through the entire retina.

As yet, neurons that function entirely without impulses have not been observed elsewhere in the vertebrate central nervous system, although invertebrates have such cells (Siegler and Burrows, 1980).

Not finding neurons with slow potentials elsewhere in the vertebrate brain does not necessarily mean that such cells are absent. Two factors may be responsible for the inability to find them: first, such cells almost certainly are small cells, which means they are difficult to record; second, most recordings from the central nervous system are extracellular and therefore do not readily allow the detection of slow-potential cells. But even if most neurons in the brain do exhibit impulse activity, local, graded potential activation of synapses can occur in distal processes of the cells. An interesting observation in this regard is that distal dendrites of large-field ganglion cells in the catfish retina make synapses (Sakai et al., 1986). In these cells, clusters of both input and output synapses were observed more than 100 μm away from the cell body—an indication of local circuit interaction in the dendritic tree of a classic, spike-firing Golgi type I neuron.

The retinal cells that function only with graded potentials have provided a unique opportunity for studying the properties of such potentials in initiating synaptic interactions. Study of the photoreceptors has revealed that exquisitely small changes in membrane potential modulate transmitter release from synaptic terminals. Potential changes of 100 μV (0.1 mV) or even less across the photoreceptor terminal membrane result in signals that reliably reach other retinal and central nervous system neurons (Fain et al., 1976). This situation is very different from the classic picture derived from the neuromuscular junction, where action potential-induced depolarizations of more than 20 mV are required in the presynaptic terminal to initiate release of neurotransmitter and synaptic activity.

We do not understand fully the mechanisms underlying the high degree of voltage sensitivity of the photoreceptor and presumably other synaptic terminals in the retina and elsewhere. As discussed in Chapter 5, part of this high sensitivity probably relates to the fact that the photoreceptors and many other retinal neurons are maintained in a highly depolarized state in the dark. They are poised, therefore, at a voltage likely to be on the steepest portion of the stimulus–release curve (see Figure 5.3), thus ensuring maximal sensitivity to small voltage changes. Other mechanisms may also be operating to provide this high sensitivity, but as yet these are unknown.

That the photoreceptors and other retinal neurons are maintained "on" in the dark and that light, for the most part, turns them "off" has also led to some revisions of our thinking about neurotransmitter action on postsynaptic cells. Again, mainly from study of the neuromuscular junction, it was believed that postsynaptic cells generally desensitize rapidly to transmitter substances. That is, over time the postsynaptic response to exogenously applied transmitter decreases substantially (Katz and Thesleff, 1957). But, many retinal cells release transmitter continually in the dark, and postsynaptic cells are polarized indefinitely in response to the impinging transmitter (Figures 5.1 and 5.2). Such findings imply that little or no desensitization can occur at these synaptic junctions. Direct tests of this hypothesis, by examining the response of cultured horizontal cells to continuously applied L-glutamate or glutamate analogues, have confirmed that no classic desensitization occurs postsynaptically in these cells (Lasater et al., 1984).

Electrical Coupling

Another mechanism of neuronal interaction, now known to occur in many parts of the vertebrate central nervous system and observed early in retinal studies, is electrical coupling between cells. Throughout the 1930s and 1940s researchers vigorously debated whether synaptic interactions occurring within the vertebrate brain were mediated primarily by chemical or electrical means (see Eccles, 1975). By 1950 unequivocal evidence for the chemical nature of synaptic transmission in the central nervous system had been established; it was then generally believed that all synaptic transmission in the vertebrate brain was chemical. In 1959 Furshpan and Potter provided the first clear evidence for electrical synaptic transmission between neurons in the crayfish. Other studies in invertebrates confirmed electrical transmission between neurons, but the extent or even the existence of such interactions in the vertebrate brain was uncertain and thought by many not to have a significant role. The retina was one of the first parts of the vertebrate brain to show clear evidence of extensive electrical coupling between cells and has provided an opportunity for studying the significance and properties of such coupling.

Electrical coupling between the receptors and between the horizontal cells in the retina has been studied most extensively, but electrical coupling between bipolar, amacrine, and perhaps even retinal ganglion cells has also been reported or suggested (Chapter 4). The function of the electrical coupling between the photoreceptors is still not well understood (see Chapter 4), but at least one major consequence of the extensive coupling between horizontal cells is to increase substantially the effective receptive field size of the cells. The processes of a single horizontal cell in fish, for example, may spread out about 100 μm, yet the measured receptive field size of one cell is several millimeters in diameter. Because horizontal cells form the antagonistic surrounds of bipolar and receptor cells, electrical coupling of the horizontal cells provides an inhibitory field much larger than that available without coupling. It is also significant that coupling between horizontal cells is specific; only cells of the same subtype are coupled. This arrangement implies that horizontal cells of the same subtype can recognize each other, and experiments with cultured horizontal cells are providing evidence that the same specificity holds under in vitro conditions as well (Lasater and Dowling, 1985a).

A particularly interesting and important finding is that the extent of electrical coupling between horizontal cells can be modified by chemical synaptic input to the cells. Dopamine applied to fish or turtle horizontal cells significantly decreases coupling between cells (Plate 3). As described in Chapter 5, this effect appears to be mediated by the second messenger, cyclic AMP, which is generated in horizontal cells by dopamine receptor activation of adenylate cyclase. Modulation of electrical synaptic activity by chemical synaptic input has not been observed elsewhere in the brain, but it seems likely that it is a mechanism of general use. The significance of this phenomenon for information processing in the fish retina, where it has been extensively studied, is that the extent and strength of surround antagonism in receptors and bipolar cells—and presumably other retinal neurons—can be modulated as a function of dopaminergic interplexiform cell activity. In the turtle the source of the dopamine that may modulate the horizontal cell coupling has not been positively identified; as yet no dopaminergic interplexiform cells have been observed in the turtle, although dopaminergic amacrine cells have been visualized. The possible significance of these observations is commented upon further in the next section.

Neurotransmitters and Neuromodulators

Although the notion of chemical transmission of signals at synapses goes back to the early part of this century (Elliot, 1904), it was not until the classic experiments of Otto Loewi in the early 1920s that this idea was taken very seriously by those interested in brain function. Loewi (1921) demonstrated that when the perfusate from around the heart of a frog whose heartbeat was slowed by vigorous electrical stimulation of the vagus nerve was collected and applied to the heart of another animal whose vagus nerve was unstimulated the heartbeat of the treated frog would also slow. He concluded that the vagus nerve, when stimulated, releases an inhibitory substance into the fluid around the heart that slows the heartbeat. Loewi and his colleagues showed later that the inhibitory substance was the same as acetylcholine, which in this particular case causes an inhibitory action.

As noted earlier in this chapter, not until the 1950s was synaptic transmission in the brain generally accepted as chemical in nature. At that time, there were few candidates for brain neurotransmitters other than acetylcholine. Researchers knew that the monoamines—epinephrine, norepinephrine, and serotonin—were present in many parts of the central nervous system; and a neuropeptide, substance P, had been extracted from brain tissue. Yet no substances that appeared primarily to mediate inhibitory synaptic interactions in the brain were then known (see Eccles, 1957).

In the 1950s and 1960s, a vigorous search for other transmitter substances took place, but progress was relatively slow. The acidic amino acids, glutamate and aspartate, were proposed as possible excitatory neurotransmitter agents, and γ-aminobutyric acid and glycine were suggested as possible inhibitory transmitters. Dopamine was also recognized as a likely agent for release at synaptic sites in the brain, but the list of chemical agents used in the vertebrate brain totaled no more than ten substances. In the 1970s, particularly as a result of the application of immunohistochemical techniques to identify specific substances within individual neurons, the number of potential neuroactive substances found in the brain increased enormously. Today forty, or even more, substances are suspected of being released from neurons in the brain. In the retina at least fifteen substances have been identified as possible neuroactive agents (Table 5.1).

At the present time little is known about the function of most proposed neuroactive substances in the brain. This is particularly true of the largest group of newly identified substances, the neuropeptides. Investigators have come to realize that the various substances act in different ways in the brain and on postsynaptic cells; so attempts at classifying their actions are being made with the hope of understanding this sudden embarrassment of riches. A simple and attractive scheme that has proved useful is a division of the agents released from synaptic sites into two general categories, neurotransmitters and neuromodulators. As described in Chapter

5, neurotransmitters mediate fast excitatory and inhibitory pathways in the brain by directly altering the membrane permeability of postsynaptic cells to one or a few ions. Neuromodulators, on the other hand, modify the activity of neurons by nonconventional means, often by the activation of biochemical mechanisms within neurons. Neuromodulators have a slow onset of action and their effects typically can last for long periods—minutes to hours, or longer.

The study of neuropharmacological mechanisms in the brain is one of the most active and rapidly developing areas in neuroscience research. Undoubtedly our ideas in this field will alter substantially over the next decade. Already, however, retinal studies may be providing key insights into what at first glance appears to be a bewildering array of substances found in, and presumably released by, synaptic terminals. For example, the evidence so far suggests that only a few substances mediate classic neurotransmitter functions in the retina: the main excitatory pathways through the retina appear to be mediated principally by an excitatory amino acid, probably L-glutamate (receptors and bipolar cells), and by acetylcholine (cholinergic amacrine cells); whereas classic inhibition in the retina appears to be mediated mainly by GABA (horizontal and amacrine cells) and by glycine (amacrine cells). That up to 80 percent of the amacrine cells in the retina take up either GABA or glycine (Pourcho, 1980), and presumably can release these agents, is strong evidence that these substances are mediating a good deal of the key inhibitory interactions within the inner plexiform layer. Blocking the action of these agents—by the infusion of specific antagonists into the retina (Caldwell et al., 1978)—dramatically alters the receptive field properties of many of the ganglion cells, evidence confirming the critical role of GABA and glycine in establishing ganglion cell receptive field organization. The glutamate analogue, 2-amino-4-phosphonobutyric acid, blocks all of the "on" activity throughout the retina and in central visual nuclei (Slaughter and Miller, 1981; Schiller, 1984), a finding indicating that just a single transmitter substance, L-glutamate, mediates the transmission of this key information from receptors to depolarizing bipolar cells. Furthermore, the excitatory amino acid antagonist, $(\pm)cis$-2, 3-piperidinedicarboxylic acid, blocks on-center ganglion cell responses (Slaughter and Miller, 1983b), a finding indicating that also from bipolar terminals a single transmitter substance is signaling all "on" activity in the inner plexiform layer.

If it is true that just a few transmitter agents are responsible for signaling the primary "on" and "off" responses and establishing the basic receptive field organizations of the retinal neurons, what are the other ten or more neuroactive substances doing in the retina? As pointed out in Chapter 5, many of these other substances are found only in a small percentage of a particular class of cells. For example, neuropeptides are present in certain amacrine cells that make up less than 1 percent of the total amacrine cell population in the bird retina (Karten and Brecha, 1982). The processes of these cells typically spread long distances (see Figure 2.3). This arrangement suggests that these cells can affect many retinal neurons, but at the same time it seems unlikely that they directly mediate high-acuity visual information. Rather, the guess is that the majority of these cells, and the substances they release, modulate the activity of neurons that give rise to primary excitatory and inhibitory signals that underlie responses and receptive field properties of the retina's output neurons, that is, the bipolar and ganglion cells.

The dopaminergic interplexiform cell in the fish may be an excellent model of a neuromodulatory neuron in the retina and brain. Although the interplexiform cells make up only a small proportion of cells in the fish's inner nuclear layer, they spread processes widely in both plexiform layers. They make abundant synapses on the horizontal cells, which are the primary inhibitory neurons in the outer plexiform layer. Dopamine has no direct effects on membrane voltage or resistance of the horizontal cell, but it activates through specific dopamine receptors the enzyme adenylate cyclase, which increases cyclic AMP levels within the cell. The cyclic AMP mediates multiple effects in the cell, including a decrease in its light responsiveness, a shrinkage of its receptive field, and a depression of its release of transmitter. All of these changes are effective ways of lessening the influence of the horizontal cells. As described in Chapter 5, dopamine and the activation of the interplexiform cells also modulates center–surround antagonism in receptor, bipolar, and presumably other retinal neurons. It is important to emphasize that dopamine and the interplexiform cell do not mediate the center–surround antagonism but rather modulate its strength. Thus, activating or blocking dopaminergic activity in the retina does not change the basic receptive field organization of the bipolar cells; instead it alters the balance of center and surround responses (Hedden and Dowling, 1978).

Little is presently known about the other possible neuromodulatory neurons and substances in the retina. There may be many such cells containing different substances, perhaps as many as a dozen in the inner plexiform layer alone. This raises the intriguing question of their role. Evidence that these neurons make specific synaptic connections in the inner plexiform layer (see Chapter 5) suggests that perhaps there is separate modulatory control of all the major excitatory and inhibitory retinal neurons and pathways. Support for this idea comes from studies on ensembles of neurons that underlie simple rhythmic behaviors in invertebrates (the so-called central pattern generators). In the spiny lobster (*Panulerus interruptus*), one well-studied ensemble comprises fourteen neurons. Only two neurotransmitters (acetylcholine and glutamate) are used by the neurons in the circuit, but four monoamines, including dopamine and serotonin, and two peptides have multiple effects on many of the neurons in the ensemble. They modulate in various ways the activity of the neurons and their firing patterns. Thus, each of these neuromodulatory agents modifies in a highly specific way the effective circuitry of the ensemble and its final output (Flamm and Harris-Warrick, 1986a, b). Further tests of such ideas are needed, and the retina appears to be an ideal part of the vertebrate central nervous system in which to determine the role of neuromodulators in brain function.

A final point concerning neuropharmacological mechanisms relates to the observation that, although horizontal cell coupling in the turtle retina is modulated by dopamine, no dopaminergic terminals are close to the horizontal cells or to their processes. A similar finding was made earlier in the fish retina, where vasoactive intestinal peptide receptors were found on horizontal cells but no vasoactive intestinal peptide-containing synaptic terminals were observed in proximity to horizontal cells (Watling and Dowling, 1983). Rather, as with dopaminergic cells in the turtle, the cells containing vasoactive intestinal peptide extend their processes only in the inner plexiform layer. These findings suggest that sometimes neuroactive substances released in the retina can have effects tens of micrometers away from the site of release, and in this sense they are acting like neurohormones (agents that diffuse and act over considerable distances in the brain). Such a situation has been found in the frog's sympathetic ganglion; there the neuropeptide leutinizing hormone releasing hormone remains active for some time after release and affects cells many micrometers away from its release site (Jan and Jan, 1982).

Receptive Field Mechanisms

Some of the most important contributions of retinal studies to our understanding of brain mechanisms come from studies on the formation of retinal receptive fields. We now understand in general terms how the center–surround organization of bipolar cells is established by the neural interactions in the outer plexiform layer. Furthermore, we are learning how more complex receptive field organizations such as direction and orientation selectivity are generated from the interplay of excitatory and inhibitory amacrine cell input to ganglion cells.

We now have quite clear evidence that the center response of the bipolar cell reflects direct receptor-to-bipolar input and that the antagonistic surround response arises from horizontal cell activity. The remaining point of debate is whether the antagonistic surround input from horizontal cells is directly onto bipolar cells or is fed back to the receptors. There is evidence for both possibilities (Chapters 3 and 4), and it may be that both mechanisms are used by different receptor systems or by different retinas. It may also be that both mechanisms are used by a single receptor system in a single species (Marshak and Dowling, 1987). In any case it appears that center–surround organization of bipolar and receptor cells can be explained by the wiring patterns between the neurons, the polarities of the synaptic interactions, and the fact that horizontal cells have a much larger receptive field than do receptor or bipolar cells.

In many respects the surround antagonism imparted to receptors and bipolar cells by the horizontal cell network resembles the lateral inhibition between the second-order eccentric cells in the horseshoe crab eye. In both cases surround illumination inhibits the central neuron's response. As with the horseshoe crab eye, the Mach band phenomenon in the vertebrate visual system (Figure 1.1) can be explained by such lateral and inhibitory interactions, and much of this phenomenon can result from synaptic interactions in the outer plexiform layer among the receptor, bipolar, and horizontal cells.

The inner plexiform layer demonstrates more complex processing of visual information; nevertheless one can suggest how such receptive fields are established (Chapter 4). Orientation-selective cells, for example, appear to be basically on- and off-center ganglion cells that receive strong antagonistic input from sustained amacrine cells. Thus, these cells appear to receive a mixed bipolar and sustained amacrine cell input (Figure 4.19). Dendritic fields of certain sustained amacrine cells in the catfish are elongated in one direction (Naka, 1980); and if such cells provide strong inhibitory input to on- and off-ganglion cells, orientation preference of the receptive fields will result (Caldwell et al., 1978). In this instance, receptive field organization can be explained on the basis of the synaptic input to the ganglion cells coupled with a specific morphology of an inhibitory neuron.

The receptive fields of direction-sensitive ganglion cells indicate that additional mechanisms operate in the inner plexiform layer and contribute to receptive field properties and organization. Most of the excitatory and inhibitory input to these cells appears to be provided by amacrine cells. The acetylcholine-releasing amacrines impart excitatory input, whereas GABA-releasing amacrine cells provide inhibition. The direction-sensitive ganglion cells respond only to the onset or offset of a light stimulus or to a moving stimulus. Thus, the input to these cells must be coupled transiently to the light stimulus; that is, they are likely to be receiving input principally from transient amacrine cells (Masland et al., 1984b). Such amacrine cells are the first neurons along the visual pathway to respond primarily in a transient way (Chapter 4). We do not understand how the transient responses of these amacrine cells are generated, but it is believed that the intrinsic membrane properties of these amacrine cells are at least partially responsible. Thus, with direction-sensitive ganglion cells, not only is cell wiring important (and not fully resolved, see Chapter 4), but also the special properties and behavior of the input neurons (and perhaps the ganglion cell also) appears crucial to the responses of the postsynaptic cells. Similar conclusions have been drawn from studies on invertebrate neurons, the neuronal responses of which reflect not only the synaptic input to the cells but also special membrane properties of the cells involved. For example, the motor neuron mediating the slow release of ink from *Aplysia* possesses a special voltage-sensitive K^+ channel in its membrane; the effect of this special channel is to oppose transiently the depolarization of the cell in response to excitatory synaptic input (Kandel, 1976). We are just beginning to uncover the membrane properties of the retinal neurons and to correlate these properties with the cell's response behavior (Tachibana, 1981; Barnes and Werblin, 1986). With newer techniques such as patch clamping of neurons and culturing of retinal cells, we will undoubtedly make rapid progress in uncovering the molecular mechanisms operating in the membranes of retinal cells and their properties.

Epilogue

Studies on the retina over the past three decades have provided much information on how this tiny bit of the brain functions, and they have helped to shape our thinking about brain mechanisms. In this chapter I have mentioned only a few of the recent findings that have contributed to our understanding of the brain. We still know relatively little of the role and function of the glial cells in the brain, and we have much more to learn about retinal glial cells. That most of the electroretinogram recorded from the intact eye or retina reflects potentials generated by the glial (Müller) cells and pigment epithelium was surprising (Chapter 6) and may have implications for understanding the origins of other field potentials recorded from the brain, such as the electroencephalogram (EEG). The processes of light and dark adaptation, outlined in Chapter 7, may seem unique to the visual system, but the neural mechanisms underlying these phenomena will undoubtedly have applicability in other areas of the brain.

Cajal recognized a century ago the great value of the retina as a "genuine neural center." He returned to it again and again in his career to answer questions about the organization of neural tissue. He notes in his autobiography (Ramón y Cajal, 1964) that "the retina has always shown itself to be generous with me." The same statement can be made by scores of investigators since, and, I am confident, by many more to come. Cajal called the retina "the oldest and most persistent of my laboratory loves." In the closing chapters of his autobiography, while discussing his long monograph devoted to the retina and optic centers of insects, he movingly summarizes his feelings for the retina and visual system. His words make a fitting close to this book.

"As the reader will remember, my devotion to the retina is ancient history. The subject always fascinated me because, to my idea, life never succeeded in constructing a machine so subtly devised and so perfectly adapted to an end as the visual apparatus. It is one of the rare cases, nevertheless, in which nature has deigned to employ physical means which are accessible to our present knowledge. I must not conceal the fact that in the study of this membrane I for the first time felt my faith in Darwinism (hypothesis of natural selection) weakened, being amazed and confounded by the supreme constructive ingenuity revealed not only in the retina and in the dioptric apparatus of the vertebrates but even in the meanest insect eye. There, in fine, I felt more profoundly than in any other subject of study the shuddering sensation of the unfathomable mystery of life."

(Ramón y Cajal, 1964, p. 576)

A P P E N D I X

R E F E R E N C E S

I N D E X

Basic Concepts and Terms in Neurobiology

Two types of cells are present in the brain: *neurons,* which receive, process, and integrate information; and *glia* (or *glial cells*), which play a supporting role for the neurons. Glial cells and their processes fill much of the space between the neurons. They insulate neuronal processes, provide a structural framework for the neurons, and help regulate levels of ions and other substances (including perhaps neurotransmitters) in the extracellular space of the brain.

Neurons are typically clustered in the brain in discrete cellular groups called *nuclei* (sing., nucleus). In some brain regions, such as the retina and the cerebellum, neuronal cell bodies (*perikarya*) are arranged in *nuclear layers.* Layers consisting mainly of processes are the *plexiform layers* (also called *molecular* or *neuropil layers*); these layers are sites of extensive interactions between neuronal cell processes.

Two basic types of neurons are often distinguished in the vertebrate brain: *Golgi type I cells,* the processes of which extend between nuclei or between plexiform layers, and *Golgi type II cells,* the processes of which are confined to a nucleus or to a single plexiform layer (Figure A.1). The Golgi type I cells are typically much larger than the Golgi type II cells. Two kinds of processes arise from

A.1 Highly simplified representation of a brain nucleus. Two types of neurons are typically present in brain nucleii: Golgi type I, whose axons project to other nuclei, and Golgi type II cells, whose processes are confined to the nucleus. The dendrites of Golgi type I cells usually only receive input, from axon terminals, but the dendrites of Golgi type II cells may both receive input and provide synaptic output.

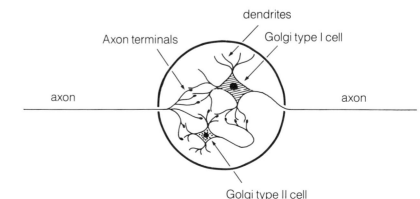

neurons: *dendrites,* which are relatively short, bushy elements and are usually tapered; and *axons,* which are longer, finer elements ending in a profusion of *axon terminals.* Neurons typically have numerous dendrites but only a single axon.

In Golgi type I cells the dendrites are specialized to receive input, and information from such cells is transmitted via the axon and its terminals. The processes of Golgi type II cells are less specialized; that is, both dendrites and axons may receive and transmit information. The principal role of the Golgi type II neuron is to integrate information within a nucleus or a plexiform layer.

Information is transmitted electrically along neurons, and nerve cells generate two basic types of potentials: graded potentials and action potentials. *Graded potentials* are generated in sensory receptors and in dendrites. They are usually sustained in nature, lasting as long as the stimulus, with an amplitude that is proportional to the strength of the stimulus. Graded potentials are local potentials (that is, they are not transmitted), and their amplitudes diminish as they move away from their site of generation. They summate (add together), and they may be either positive or negative. Graded potentials have no threshold; for example, a single quantum of light absorbed by a photoreceptor generates a small, graded potential.

Action potentials, often called *impulses* or *spikes,* are relatively large (0.1 V), transient ($\sim$1–2 msec), all-or-none potentials that are generated along axons. They signal strength of stimulus by frequency rather than by amplitude. Action potentials are usually transmitted, that is, they are continually regenerated along an axon; hence, an action potential at the end of an axon is identical in amplitude to one at the beginning. Action potentials also have a *threshold;* for example, for an action potential to be generated, voltage across the cell membrane (*membrane potential*) must be changed by about 15 mV.

Figure A.2 illustrates the interplay between graded and action potentials in the nervous system. Some cells with short axons may function with graded potentials only (as illustrated by the sensory receptor cell in Figure A.2). Some Golgi type II neurons have no axons (these cells are called *axonless,* or *amacrine, cells*) and also may function without action potentials. Most neurons, however, have both kinds of potentials. The graded potentials in the dendrites summate to change the membrane potential of the cell by a sufficient amount to trigger the firing of action potentials in the axons. Thus, graded potentials generate action potentials in a neuron and, for this reason, are sometimes called *generator potentials.*

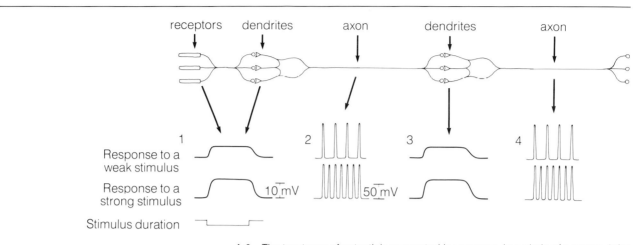

receptors   dendrites   axon   dendrites   axon

1

Response to a
weak stimulus

Response to a
strong stimulus   10 mV   50 mV

Stimulus duration

A.2   The two types of potentials generated by neurons. In a chain of neurons, information is carried along axons by action potentials. At the junctions between cells (synapses), graded potentials are generated in the dendrites of the postsynaptic (receiving) neurons. The graded potentials in the dendrites trigger the firing of action potentials in the postsynaptic cell's axon, and information is thus transmitted to the cell's axon terminals, where it is passed on to other neurons. 1 and 3, graded potentials; 2 and 4, action potentials.

Underlying both graded and action potentials is a *resting potential,* which is the membrane potential of both unexcited neurons and other cells. In neurons, the resting potential is normally about -70 mV; the cell's inside charge is negative relative to the outside charge. The resting potential across cell membranes is established by a differential distribution of ions across the membrane (that is, on the two sides of the membrane) and a differential permeability of the membrane to ions (Figure A.3). Unequal ion distribution across cell membranes is ultimately the result of the activity of *active ion pumps* in the cell membrane. These pumps move ions across the membrane and "up" (or "against") their concentration gradient (that is, from a region of low ion concentration to one of higher ion concentration). Differential *permeability* of the cell membrane to ions is due to *selective ion channels* that lie within the cell membrane and let only certain ions cross the membrane. In most cases, potassium ion (K^+) concentration is high intracellularly, whereas sodium ion (Na^+) and chloride ion (Cl^-) concentrations are high extracellularly. At rest, nerve cell membranes are most permeable to K^+, less so to Cl^-, and least permeable to Na^+.

Graded potentials are generated in a receptor or dendrite by a

A.3   Schematic representation of the relative permeability of a neuronal membrane to various ions at rest and during the rising (early) and falling (late) phases of the action potential.

At rest, the membrane is very impermeable to Na$^+$ (indicated by the steep barrier Na$^+$ must traverse to enter the cell). The membrane is most permeable to K$^+$ (indicated by the modest barrier K$^+$ must overcome to traverse the membrane). Cl$^-$ permeability of the membrane is intermediate to that of Na$^+$ and K$^+$, and the membrane is completely impermeable to organic anions (An$^-$). The relative concentrations of the ions on either side of the membrane are indicated by the size of the letters depicting the ions. Thus, high concentrations of Na$^+$ and Cl$^-$ are found outside the cell membrane, and high concentrations of K$^+$ and An$^-$ are found inside the cell.

At rest, the inside of the cell is more negative than the outside of the cell (by 70 mV), primarily because the membrane is relatively permeable to K$^+$, thereby allowing some K$^+$ to cross the cell membrane. This K$^+$ movement leaves excess ($-$) charges inside the cell and adds ($+$) charges to the outside of the cell.

During the generation of an action potential, the membrane becomes very permeable to Na$^+$, thereby allowing a rapid buildup of ($+$) charge within the cell and causing membrane potential to change from -70 mV to $+50$ mV (see Figure A.5). After a short time ($\sim$1 msec), the membrane once again becomes impermeable to Na$^+$ ions, but now the membrane becomes more permeable to K$^+$ ions. The rapid efflux of K$^+$ across the membrane removes the positive charge from inside the cell and restores the resting potential. During the generation of an action potential, membrane permeabilities to Cl$^-$ and An$^-$ do not change.

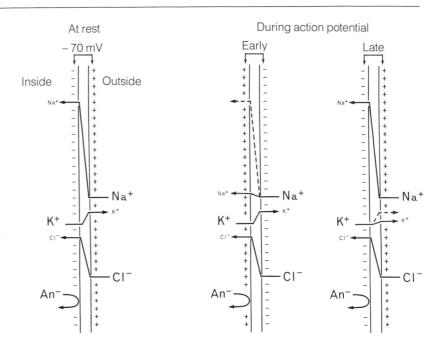

At rest

During action potential

stimulus that changes membrane permeability by altering specific ion channels in the membrane so that one or a combination of ions can pass more freely across the cell membrane. In other words, after stimulation ions can move "down" their concentration gradients (that is, move from a region of high concentration to one of lower concentration). This ion movement customarily changes membrane potential. The stronger the stimulus, the more ions crossing the cell membrane and the larger the change in potential. Thus, the response is graded with (proportional to) stimulus strength.

The voltage developed across the membrane depends not only on the number of ions flowing (current, or I) but also on the resistance (R) of the membrane, in accord with Ohm's law, $V = IR$. Membrane resistance reflects the ease with which ions can traverse the membrane. As membrane permeability changes, so usually does membrane resistance. Membrane conductance (G), a quantity closely related to membrane permeability, is the inverse of membrane resistance, $G = 1/R$; it reflects not only the ionic permeability of the membrane but also the numbers and distribution of the permeant ions. An easy way to detect changes in membrane permeability (conductance) is to measure membrane resistance. This can be done by passing a known current pulse into a neuron and measuring the voltage that develops across the membrane. The resistance is determined by dividing the voltage by the current, $R = V/I$.

Axons contain ion channels whose permeabilities depend on voltage across the membrane (called voltage-dependent channels). When the membrane is *depolarized* (that is, when the membrane potential becomes less negative), Na^+ channels begin to open. Action potentials are generated when the membrane potential is decreased by about 15mV: from -70 to -55 mV. At threshold ($\sim$ -55 mV), a large and rapid change in membrane permeability to Na^+ is initiated (Figure A.3). This change allows enough positive Na^+ ions into the cell to shift membrane potential from -55 mV to $+50$ mV. The increased permeability to Na^+ is transient, lasting only about 1 msec. Then K^+ channels open, thereby allowing K^+ to leave the cell (Figure A.3). The inside of the cell therefore becomes more negative, and the membrane potential is quickly restored to -70 mV. All of this occurs in 1–2 msec, shortly after which a new action potential can be generated in the axon. The final step involves the membrane pumps, which extrude Na^+ from the cell and pump in K^+ to restore the concentration gradients across the membrane. However, so few ions cross the membrane during the generation of an action potential, relative to the numbers of ions inside and outside of nerve cells, that it takes hundreds to thousands of action potentials to alter the concentration gradient significantly. Thus, the pumps can act relatively slowly. It should be noted, finally, that action potentials that depend on Ca^{2+} are found in some nerve cells.

Nerve and sensory receptor cells pass on information at *synapses* (Figure A.4), which are sites at which these cells make functional contact. Most synapses in the brain are chemical, but electrical synapses are also known. *Chemical synapses* release substances that diffuse across a small extracellular space (*synaptic cleft,* usually 20–30 nm) and interact with specific receptor proteins on the receptive cell. The chemicals released at synapses are called *neuroactive substances,* more specifically *neurotransmitters* or *neuromodulators.* They are released at specialized sites that are present most commonly in axon terminals but also are present in some dendrites (particularly of Golgi type II cells). The site of release is the *presynaptic membrane;* the site where a neuroactive substance acts is the *postsynaptic membrane.* At electrical synapses, the membranes of the interacting cells come into close apposition (about 2

A.4 Typical excitatory and inhibitory synapses, as observed by electron microscopy. Excitatory synapses are characterized by round synaptic vesicles; inhibitory synapses often have flattened vesicles. Membrane specializations, along with a cluster of vesicles, are characteristically seen at sites of transmitter release. At excitatory synapses a small depolarizing (positive) potential is elicited in the postsynaptic cell and decreases membrane potential by 5 mV or so from resting membrane potential (-70 mV); at most inhibitory synapses a small hyperpolarizing (negative) potential is generated, thereby making the membrane potential of the neuron more negative. Acetylcholine and glutamate are typical excitatory transmitters; γ-aminobutyric acid and glycine are inhibitory transmitters.

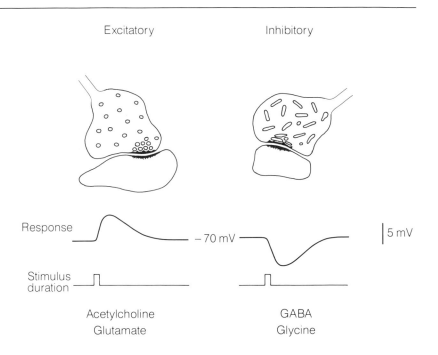

nm), forming *gap junctions*. Relatively nonselective ion channels extend from one cell to another at gap junctions, thereby allowing ions and small molecules to pass relatively freely between the two cells in either direction. Thus, potential changes in one cell are transmitted rapidly and faithfully to the adjacent cell via such junctions.

In the cytoplasm adjacent to the presynaptic membrane of the chemical synapse, and often clustered on the membrane itself, are small, membrane-bounded *synaptic vesicles* that store neuroactive substances (Figure A.4). Specializations on the presynaptic membrane (*electron-dense projections*) or in the adjacent cytoplasm (*synaptic ribbons*) help bind the vesicles to the membrane, from which position they can release their contents into the synaptic cleft. Release occurs when the presynaptic membrane is depolarized, but release generally requires Ca^{2+}. In short, depolarization opens specific (voltage-sensitive) Ca^{2+} channels that are near or in the presynaptic membrane. The influx of Ca^{2+} binds vesicles to the membrane, thereby enabling the vesicles to release neuroactive substance.

At many synapses, the substances released (called *neurotransmitters* in these cases) interact with specific receptors that are lo-

A.5 Schematic diagrams showing the interplay of excitatory and inhibitory synaptic inputs to a neuron. The record in **b** is typical of an intracellular recording made with a micropipette inserted in the cell body of a neuron (**a**). In this location potentials generated in the dendrites, cell body, and axon are detected.

When the pipette is inserted into the neuron, a resting potential of − 70 mV is recorded. Excitatory input (depicted by tics along the abscissa E) causes depolarizing graded (synaptic) potentials, which summate and bring the membrane potential of the cell to action potential threshold (∼ − 55 mV). Generation of an action potential causes a very rapid reversal of membrane potential, followed by rapid recovery (see Figure A.3). With continued stimulation of the excitatory input, the cell again depolarizes and additional action potentials are fired. Inhibitory input to the cell (depicted by tics along the abscissa I) causes the generation of hyperpolarizing graded (synaptic) potentials, which negate the effects of the excitatory synaptic potentials and drive the membrane potential away from spike firing threshold. The cell does not fire while the inhibitory input is active, even though the excitatory input continues. When the inhibitory input is terminated, the cell once again depolarizes and fires action potentials.

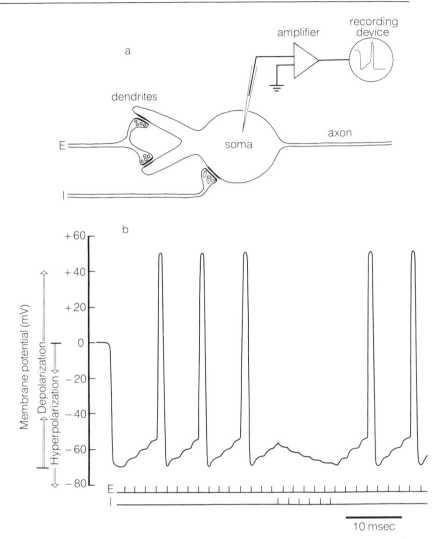

cated on the postsynaptic membrane and are linked to ion channels (often called chemical-sensitive channels). Activation of these receptors changes membrane permeability and generates graded potentials in the postsynaptic cell (Figure A.4). If the changes in membrane permeability yield a net increase of positive charge in the postsynaptic cell (by influx of Na⁺ into the cell, for example), the cell depolarizes and the graded potential is *excitatory* (Figure A.4). That is, membrane potential moves toward action potential threshold (Figure A.5), promotes release of neuroactive substances from nearby synapses, or both. If membrane permeability changes yield a net increase of negative charges in the cell (for instance, by Cl⁻

influx or K$^+$ efflux or both), the cell *hyperpolarizes*. In this case the synaptic potential is *inhibitory;* that is, it moves membrane potential away from action potential threshold (Figure A.5) and decreases release of neuroactive substances from nearby synapses.

Different neurotransmitter agents are usually released at excitatory and inhibitory synapses. *Acetylcholine* and the acidic amino acids *L-glutamate* are typical excitatory neurotransmitters, whereas *γ-aminobutyric acid* (GABA) and *glycine* are typical inhibitory transmitters. At many excitatory synapses the synaptic vesicles are round; at many inhibitory synapses the vesicles are flattened. This difference may reflect the fact that different neurotransmitter agents are released at most excitatory and inhibitory synapses.

Not all substances released from vesicles at synaptic sites act on postsynaptic cells as just described. A second class of neuroactive substances (*neuromodulators*) is often distinguished. Neuromodulators modify the activity of postsynaptic neurons in a variety of ways, but they usually do not directly alter membrane permeability. For example, neuromodulators interact with nerve cell receptors that are linked to membrane-bound or intracellular enzymes. When these receptors are excited, the enzymes are activated and bring about various biochemical changes in the postsynaptic neuron. Neuromodulators typically have a slow onset of action, and their effects can last for minutes, hours, days, or longer.

Figure A.6 shows how a neuromodulatory system works. The

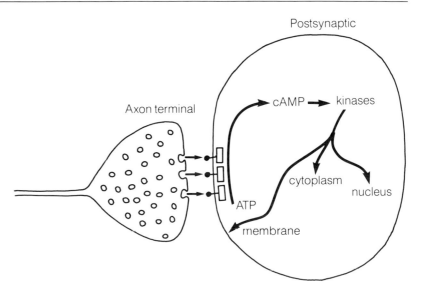

A.6   Effects of a neuromodulator. The neuromodulator, released at a synapse, interacts with receptors on a postsynaptic neuron. Receptor stimulation activates intracellular enzyme systems, in this case an enzyme (adenylate cyclase) that converts ATP to cyclic AMP. Cyclic AMP is known to activate other enzymes called kinases, which add or remove phosphate groups from molecules and thereby activate or inactivate them. Cyclic AMP-dependent kinases may exert their effects in the nucleus, in the cytoplasm, or at the cell membrane, thus potentially altering a variety of cellular processes.

neuromodulatory substance interacts with the postsynaptic receptor and activates the enzyme *adenylate cyclase,* which converts adenosine triphosphate *(ATP)* to *cyclic AMP.* Cyclic AMP, called a second messenger, in turn activates other enzymes, called *kinases,* which catalyze reactions that add phosphate groups to cell molecules. By this process, called phosphorylation, cells turn on or turn off biochemical processes. Such kinases can act at many different levels within the cell, from the nucleus to the cell membrane. Enzymes other than adenylate cyclase are also known to be activated by neuromodulators. Some of these enzymes produce other second messengers that interact with other kinases or other molecules. Neuroactive substances that act primarily as neuromodulators include the *monoamines: dopamine, norepinephrine,* and *serotonin* and a number of the *neuropeptides.*

References

Aguilar, M., and W. S. Stiles. 1954. Saturation of the rod mechanism of retina at high levels of stimulation. *Opt. Acta* 1:59–65.

Allen, R. A. 1969. The retinal bipolar cells and their synapses in the inner plexiform layer. In *The Retina: Morphology, Functional and Clinical Characteristics*, vol. 8, ed. B. R. Straatsma, M. O. Hall, R. A. Allen, and F. Crescitelli. Los Angeles: University of California Press, pp. 101–143.

Ames, A., III, and D. A. Pollen. 1969. Neurotransmission in central nervous tissue: a study of isolated rabbit retina. *J. Neurophysiol.* 32:424–442.

Amthor, F. R., C. W. Oyster, and E. Takahashi. 1984. Morphology of on–off direction-selective ganglion cells in the rabbit retina. *Brain Res.* 298:187–190.

Ariel, M., and N. W. Daw. 1982a. Effects of cholinergic drugs on receptive field properties of rabbit retinal ganglion cells. *J. Physiol.* 324:135–160.

———— 1982b. Pharmacological analysis of directionally sensitive rabbit retinal ganglion cells. *J. Physiol.* 324:161–185.

Ariel, M., E. M. Lasater, S. C. Mangel, and J. E. Dowling. 1984. On the sensitivity of H1 horizontal cells of the carp retina to glutamate, aspartate, and their agonists. *Brain Res.* 295:179–183.

Armett-Kibel, C., I. A. Meinertzhagen, and J. E. Dowling. 1977. Cellular and synaptic organization in the lamina of the dragon-fly, *Sympetrum rubicundulum*. *Proc. R. Soc. Lond.* B 196:385–413.

Armington, J. C. 1974. *The Electroretinogram*. New York: Academic Press.

Ashmore, J. F., and G. Falk. 1980. Responses of rod bipolar cells in the dark-adapted retina of dogfish, *Scyliorhinus canicula*. *J. Physiol.* 300:115–150.

Attwell, D., and M. Wilson. 1980. Behavior of the rod network in the tiger salamander retina mediated by membrane properties of individual rods. *J. Physiol.* 309:287–315.

Attwell, D., M. Wilson, and S. M. Wu. 1984. A quantitative analysis of interactions between photoreceptors in the salamander (*Ambystoma*) retina. *J. Physiol.* 352:703–737.

Ayoub, G. S., and D. M. K. Lam. 1984. The release of γ-aminobutyric acid from horizontal cells of the goldfish (*Carassius auratus*) retina. *J. Physiol.* 355:191–214.

Bader, C. R., D. Bertrand, and E. A. Schwartz. 1982. Voltage-activated and calcium-activated currents studied in solitary rod inner segments from the salamander retina. *J. Physiol.* 331:253–284.

Bailey, C. H., and M. Chen. 1983. Morphological basis of long-term habituation and sensitization in *Aplysia*. *Science* 220:91–93.

Barlow, H. 1953. Summation and inhibition in the frog's retina. *J. Physiol.* 119:69–88.

Barlow, H. B., R. FitzHugh, and S. W. Kuffler. 1957. Change of organization of the receptive fields of the cat's retina during dark adaptation. *J. Physiol.* 137:338–354.

Barlow, H. B., R. M. Hill, and W. R. Levick. 1964. Retinal ganglion cells responding selectively to direction and speed of image motion in the rabbit. *J. Physiol.* 173:377–407.

Barlow, H. B., and W. R. Levick. 1965. The mechanism of directionally selective units in the rabbit's retina. *J. Physiol.* 178:477–504.

Barlow, R. B., Jr., and D. A. Quarles, Jr. 1975. Mach bands in the lateral eye of *Limulus*: comparison of theory and experiment. *J. Gen. Physiol.* 65:709–730.

Barnes, S., and F. Werblin. 1986. Gated currents generate single spike activity in amacrine cells of the tiger salamander retina. *Proc. Natl. Acad. Sci. USA* 83:1509–1512.

Bastian, B. L., and G. L. Fain. 1979. Light adaptation in toad rods: requirement for an internal messenger which is not calcium. *J. Physiol.* 297:493–520.

Baughman, R. W., and C. R. Bader. 1977. Biochemical characterization and cellular localization of the cholinergic system in the chicken. *Brain Res.* 138:469–485.

Baylor, D. A. 1987. Photoreceptor signals and vision. *Invest. Ophthalmol. Vis. Sci.* 28:34–49.

Baylor, D. A., and M. G. F. Fuortes. 1970. Electrical responses of single cones in the retina of the turtle. *J. Physiol.* 207:77–92.

Baylor, D. A., M. G. F. Fuortes, and P. M. O'Bryan. 1971. Receptive fields of single cones in the retina of the turtle. *J. Physiol.* 214:256–294.

Baylor, D. A., and A. L. Hodgkin. 1973. Detection and resolution of visual stimuli by turtle photoreceptors. *J. Physiol.* 234:163–198.

——— 1974. Changes in time scale and sensitivity in turtle photoreceptors. *J. Physiol.* 242:729–758.

Baylor, D. A., A. L. Hodgkin, and T. D. Lamb. 1974. The electrical response of turtle cones to flashes and steps of light. *J. Physiol.* 242:685–728.

Baylor, D. A., T. D. Lamb, and K.-W. Yau. 1979a. The membrane current of single rod outer segments. *J. Physiol.* 288:589–611.

——— 1979b. Responses of retinal rods to single photons. *J. Physiol.* 288:613–634.

Baylor, D. A., and B. J. Nunn. 1986. Electrical properties of the light-sensitive conductance of rods of the salamander *Ambystoma tigrinum*. *J. Physiol.* 371:115–145.

Belgum, J. H., D. R. Dvorak, and J. S. McReynolds. 1982. Sustained synaptic input to ganglion cells of mudpuppy retina. *J. Physiol.* 326:91–108.

Bennett, M. V. L. 1970. Comparative physiology: electric organs. *Annu. Rev. Physiol.* 32:471–528.

Bennett, M. V. L., E. Aljure, Y. Nakajima, and G. D. Pappas. 1963. Electrotonic junctions between teleost spinal neurons: electrophysiology and ultrastructure. *Science* 141:262–264.

Bloomfield, S. A., and J. E. Dowling. 1985a. Roles of aspartate and glutamate in synaptic transmission in rabbit retina. I. Outer plexiform layer. *J. Neurophysiol.* 53:699–713.

——— 1985b. Roles of aspartate and glutamate in synaptic transmission in rabbit retina. II. Inner plexiform layer. *J. Neurophysiol.* 53:714–725.

Bolz, J., T. Frumkes, T. Voigt, and H. Wässle. 1985a. Action and localization of γ-aminobutyric acid in the cat retina. *J. Physiol.* 362:369–393.

Bolz, J., P. Thier, T. Voigt, and H. Wässle. 1985b. Action and localization of glycine and taurine in the cat retina. *J. Physiol.* 362:395–413.

Bortoff, A. 1964. Localization of slow potential responses in the *Necturus* retina. *Vision Res.* 4:626–627.

Boycott, B. B., and J. E. Dowling. 1969. Organization of the primate retina: light microscopy. *Philos. Trans. R. Soc. Lond. B* 255:109–184.

Boycott, B. B., J. E. Dowling, S. K. Fisher, H. Kolb, and A. M. Laties. 1975. Interplexiform cells of the mammalian retina and their comparison with catecholamine-containing retinal cells. *Proc. R. Soc. Lond. B* 191:353–368.

Boycott, B. B., and J. M. Hopkins. 1981. Microglia in the retina of monkey and other mammals: its distinction from other types of glia and horizontal cells. *Neuroscience* 6:679–688.

Boycott, B. B., and H. Kolb. 1973. The connections between bipolar cells and photoreceptors in the retina of the domestic cat. *J. Comp. Neurol.* 148:115–140.

Boycott, B. B., L. Peichl, and H. Wässle. 1978. Morphological types of horizontal cell in the retina of the domestic cat. *Proc. R. Soc. Lond. B* 203:229–245.

Boycott, B. B., and H. Wässle. 1974. The morphological types of ganglion cells of the domestic cat's retina. *J. Physiol.* 240:397–419.

Boynton, R. M., and D. N. Whitten. 1970. Visual adaptation in monkey cones: recording of late receptor potentials. *Science* 170:1423–1426.

Brecha, N. C., W. Eldred, R. O. Kuljis, and H. J. Karten. 1984. Identification and localization of biologically active peptides in the vertebrate retina. *Prog. Ret. Res.* 3:185–226.

Brecha, N., H. J. Karten, and C. Laverack. 1979. Enkephalin-containing amacrine cells in the avian retina: immunohistochemical localization. *Proc. Natl. Acad. Sci. USA* 76:3010–3014.

Brin, K. P., and H. Ripps. 1977. Rhodopsin photoproducts and rod sensitivity in the skate retina. *J. Gen. Physiol.* 69:97–120.

Brindley, G. S. 1960. *Physiology of the Retina and the Visual Pathway*. London: Edward Arnold.

Brindley, G. S., and A. R. Gardner-Medwin. 1966. The origin of the early receptor potential of the retina. *J. Physiol.* 182:185–194.

Brown, J. H., and M. H. Makman. 1972. Stimulation by dopamine of adenylate cyclase in retinal homogenates and of adenosine 3′,5′-cyclic monophosphate formation in intact retina. *Proc. Natl. Acad. Sci. USA* 69:539–543.

Brown, K. T. 1968. The electroretinogram: its components and their origins. *Vision Res.* 8:633–677.

Brown, K. T., and M. Murakami. 1964. A new receptor potential of the monkey retina with no detectable latency. *Nature* 201:626–628.

Brown, K. T., and T. N. Wiesel. 1961. Analysis of the intraretinal electroretinogram in the intact cat eye. *J. Physiol.* 158:229–256.

Brown, P. K. 1972. Rhodopsin rotates in the visual receptor membrane. *Nature* 236:35–38.

Brown, P. K., I. R. Gibbons, and G. Wald. 1963. The visual cells and visual pigment of the mudpuppy *Necturus*. *J. Cell Biol.* 19:79–106.

Brown, P. K., and G. Wald. 1956. The neo-b isomer of vitamin A and retinene. *J. Biol. Chem.* 222:865–877.

——— 1963. Visual pigments in human and monkey retinas. *Nature* 200:37–43.

——— 1964. Visual pigments in single rods and cones of the human retina. *Science* 144:45–51.

Brunken, W. J., and N. W. Daw. 1986. 5-HT$_2$ antagonists reduce ON responses in the rabbit retina. *Brain Res.* 384:161–165.

Brunken, W. J., P. Witkovsky, and H. J. Karten. 1986. Retinal neurochemistry of three elasmobranch species: an immunohistochemical approach. *J. Comp. Neurol.* 243:1–12.

Bruun, A., B. Ehinger, and K. Tornqvist. 1986. NPY immunoreactive neurons in the vertebrate retina. In *Retinal Signal Systems, Degeneration and Transplants*, vol. 9, ed. E. Agardh and B. Ehinger. Amsterdam: Elsevier, pp. 89–103.

Burkhardt, D. A. 1970. Proximal negative response of frog retina. *J. Neurophysiol.* 33:405–420.

——— 1972. Effects of picrotoxin and strychnine upon electrical activity in the proximal retina. *Brain Res.* 43:246–249.

Burkhardt, D. A., and P. Whittle. 1973. Intensity coding in the frog retina: quantitative relations between impulse and graded activity. *J. Gen. Physiol.* 61:305–322.

Byzov, A. L. 1965. Functional properties of different cells in the retina of cold-blooded vertebrates. *Cold Spring Harbor Symp. Quant. Biol.* 30:547–558.

Byzov, A. L., and J. A. Trifonov. 1968. The response to electric stimulation of horizontal cells in the carp retina. *Vision Res.* 8:817–822.

Caldwell, J. H., and N. W. Daw. 1978a. New properties of rabbit retinal ganglion cells. *J. Physiol.* 276:257–276.

——— 1978b. Effects of picrotoxin and strychnine on rabbit retinal ganglion cells: changes in centre surround receptive fields. *J. Physiol.* 276:299–310.

Caldwell, J. H., N. W. Daw, and H. J. Wyatt. 1978. Effects of picrotoxin and strychnine on rabbit retinal ganglion cells: lateral interactions for cells with more complex receptive fields. *J. Physiol.* 276:277–298.

Cervetto, L., and E. F. MacNichol, Jr. 1972. Inactivation of horizontal cells in turtle retina by glutamate and aspartate. *Science* 178:767–768.

Cha, J. H., D. O'Brien, and J. E. Dowling. 1986. Effects of D-aspartate on excitatory amino acid-induced release of [^{3}H]GABA from goldfish retina. *Brain Res.* 376:140–148.

Chan, R. Y. and K.-I. Naka. 1976. The amacrine cell. *Vision Res.* 16:1119–1129.

Cleland, B. G., M. W. Dubin, and W. R. Levick. 1971. Sustained and transient neurons in the cat's retina and lateral geniculate nucleus. *J. Physiol.* 217:473–496.

Cleland, B. G., and W. R. Levick. 1974a. Brisk and sluggish concentrically organized ganglion cells in the cat's retina. *J. Physiol.* 240:421–456.

——— 1974b. Properties of rarely encountered types of ganglion cells in the cat's retina and an overall classification. *J. Physiol.* 240:457–492.

Cobb, W., and H. Morton. 1952. The human retinogram in response to high-intensity flashes. *Electroencephalogr. Clin. Neurophysiol.* 4:547–556.

Cohen, A. I. 1963. Vertebrate retinal cells and their organization. *Biol. Rev.* 38:427–459.

——— 1965. Some electron microscopic observations on inter-receptor contacts in the human and macaque retinae. *J. Anat.* 99:595–610.

Cohen, E., and P. Sterling. 1986. Accumulation of [³H] glycine by cone bipolar neurons in the cat retina. *J. Comp. Neurol.* 250:1–7.

Cohen, J. E., and J. E. Dowling. 1983. The role of the retinal interplexiform cell: effects of 6-hydroxydopamine on the spatial properties of carp horizontal cells. *Brain Res.* 264:307–310.

Cone, R. A. 1963. Quantum relations of the rat electroretinogram. *J. Gen. Physiol.* 46:1267–1286.

——— 1964. The early receptor potential of the vertebrate retina. *Nature* 204:736–739.

——— 1967. Early receptor potential: photoreversible charge displacement in rhodopsin. *Science* 155:1128–1131.

——— 1972. Rotational diffusion of rhodopsin in the visual receptor membrane. *Nature* 236:39–43.

Cone, R. A., and W. H. Cobbs, III. 1969. The rhodopsin cycle in the living eye of the rat. *Nature* 221:820–822.

Cornwall, M. C., A. Fein, and E. F. MacNichol, Jr. 1983. Spatial localization of bleaching adaptation in isolated vertebrate rod photoreceptors. *Proc. Natl. Acad. Sci. USA* 80:2785–2788.

Cowan, W. M., and T. P. S. Powell. 1963. Centrifugal fibers in the avian visual system. *Proc. R. Soc. Lond. B* 158:232–252.

Craik, K. J. W., and M. D. Vernon. 1941. The nature of dark adaptation. *Br. J. Psychol.* 32:62–81.

Crawford, B. H. 1937. The change of visual sensitivity with time. *Proc. R. Soc. Lond. B* 123:69–89.

Custer, N. V. 1973. Structurally specialized contacts between the photoreceptors of the retina of the axolotl. *J. Comp. Neurol.* 151:35–56.

Dacheux, R. F. 1982. Connections of the small bipolar cells with the photoreceptors in the turtle. An electron microscope study of Golgi-impregnated, gold-toned retinas. *J. Comp. Neurol.* 205:55–62.

Dacheux, R. F., and R. F. Miller. 1976. Photoreceptor-bipolar cell transmission in the perfused retina eyecup of the mudpuppy. *Science* 191:963–964.

——— 1981. An intracellular electrophysiological study of the ontogeny of functional synapses in the rabbit retina. 1. Receptors, horizontal and bipolar cells. *J. Comp. Neurol.* 198:307–326.

Dacheux, R. F., and E. Raviola. 1986. The rod pathway in the rabbit retina: a depolarizing bipolar and amacrine cell. *J. Neurosci.* 6:331–345.

Da Prada, M. 1977. Dopamine content and synthesis in retina and *N. accumbens septi:* pharmacological and light-induced modifications. In *Advances in Biochemical Psychopharmacology,* vol. 16, *Nonstriatal Dopaminergic Neurons,* ed. E. Costa and G. L. Gessa. New York: Raven Press, pp. 311–319.

Davis, G. W., and K.-I. Naka. 1980. Spatial organizations of catfish retinal neurons. I. Single and random bar stimulation. *J. Neurophysiol.* 43:807–831.

Daw, N. 1973. Neurophysiology of color vision. *Physiol. Rev.* 53:571–611.

Del Castillo, J., and B. Katz. 1954. The effects of magnesium on the activity of motor nerve endings. *J. Physiol.* 124:553–559.

Detwiler, P. B., A. L. Hodgkin, and P. A. McNaughton. 1980. Temporal and spatial characteristics of the voltage response of rods in the retina of the snapping turtle. *J. Physiol.* 300:213–250.

Dewar, J., and J. G. McKendrick. 1873. On the physiological action of light. *Trans. R. Soc. Edinburgh* 27:141–166.

Dick, E., and R. F. Miller. 1978. Light-evoked potassium activity in mudpuppy retina: its relationship to the b-wave of the electroretinogram. *Brain Res.* 154:388–394.

——— 1981. Peptides influence retinal ganglion cells. *Neurosci. Lett.* 26:131–135.

Djamgoz, M. B. A., W. K. Stell, C. A. Chin, and D. M.-K. Lam. 1981. An opiate system in the goldfish retina. *Nature* 292:620–623.

Donner, K. O., and T. Reuter. 1965. The dark-adaptation of single units in the frog's retina and its relation to the regeneration of rhodopsin. *Vision Res.* 5:615–632.

——— 1968. Visual adaptation of the rhodopsin rods in the frog's retina. *J. Physiol.* 199:59–87.

Dowling, J. E. 1960. The chemistry of visual adaptation in the rat. *Nature* 188:114–118.

——— 1963. Neural and photochemical mechanisms of visual adaptation in the rat. *J. Gen. Physiol.* 46:459–474.

——— 1967. The organization of vertebrate visual receptors. In *Molecular Organization and Biological Func-*

tion, ed. J. M. Allen. New York: Harper and Row, pp. 186–212.

——— 1968. Synaptic organization of the frog retina: an electron microscopic analysis comparing the retinas of frogs and primates. *Proc. R. Soc. Lond. B* 170:205–228.

——— 1970. Organization of vertebrate retinas. *Invest. Ophthalmol.* 9:655–680.

——— 1974. Synaptic arrangements in the vertebrate retina: the photoreceptor synapse. In *Synaptic Transmission and Neuronal Interaction*, ed. M. V. L. Bennett. New York: Raven Press, pp. 87–101.

——— 1977. Receptoral and network mechanisms of visual adaptation. *Neurosci. Res. Prog. Bull.* 15:1–12.

——— 1979. Information processing by local circuits: the vertebrate retina as a model system. In *The Neurosciences: Fourth Study Program*, ed. F. O. Schmitt and F. G. Worden. Cambridge, Mass.: MIT Press, pp. 213–216.

Dowling, J. E., and B. B. Boycott. 1966. Organization of the primate retina: electron microscopy. *Proc. R. Soc. Lond. B* 166:80–111.

Dowling, J. E., J. E. Brown, and D. Major. 1966. Synapses of horizontal cells in rabbit and cat retinas. *Science* 153:1639–1641.

Dowling, J. E., and R. L. Chappell. 1972. Neural organization of the median ocellus of the dragonfly. II. Synaptic structure. *J. Gen. Physiol.* 60:148–165.

Dowling, J. E., and W. M. Cowan. 1966. An electron microscope study of normal and degenerating centrifugal fiber terminals in the pigeon retina. *Z. Zellforsch. Mikrosk. Anat.* 71:14–28.

Dowling, J. E., and M. W. Dubin. 1984. The vertebrate retina. In *Handbook of Physiology*, vol. 2, pt. 1, ed. S. R. Geiger. Baltimore: American Physiological Society, pp. 317–340.

Dowling, J. E., and B. Ehinger. 1975. Synaptic organization of the amine-containing interplexiform cells of the goldfish and Cebus monkey retinas. *Science* 188:270–273.

——— 1978a. The interplexiform cell system. I. Synapses of the dopaminergic neurons of the goldfish retina. *Proc. R. Soc. Lond. B* 201:7–26.

——— 1978b. Synaptic organization of the dopaminergic neurons in the rabbit retina. *J. Comp. Neurol.* 180:203–220.

Dowling, J. E., B. Ehinger, and I. Floren. 1980. Fluorescence and electron microscopical observations on the amine-accumulating neurons of the Cebus monkey retina. *J. Comp. Neurol.* 192:665–685.

Dowling, J. E., E. M. Lasater, R. Van Buskirk, and K. J. Watling. 1983. Pharmacological properties of isolated fish horizontal cells. *Vision Res.* 23:421–432.

Dowling, J. E., and H. Ripps. 1970. Visual adaptation in the retina of the skate. *J. Gen. Physiol.* 56:491–520.

——— 1971. S-Potentials in the skate retina: intracellular recordings during light and dark adaptation. *J. Gen. Physiol.* 58:163–189.

——— 1972. Adaptation in skate photoreceptors. *J. Gen. Physiol.* 60:698–719.

——— 1973. Neurotransmission in the distal retina: the effect of magnesium on horizontal cell activity. *Nature* 242:101–103.

——— 1976a. Potassium and retinal sensitivity. *Brain Res.* 107:617–622.

——— 1976b. From sea to sight. *Oceanus* 19:28–33.

——— 1977. The proximal negative response and visual adaptation in the skate retina. *J. Gen. Physiol.* 69:57–74.

Dowling, J. E., and G. Wald. 1958. Vitamin A deficiency and night blindness. *Proc. Natl. Acad. Sci. USA* 44:648–661.

——— 1960. The biological activity of vitamin A acid. *Proc. Natl. Acad. Sci. USA* 46:587–608.

Dowling, J. E., and K. J. Watling. 1981. Dopaminergic mechanisms in the teleost retina. II. Factors affecting the accumulation of cyclic AMP in pieces of intact carp retina. *J. Neurochem.* 36:569–579.

Dowling, J. E., and F. S. Werblin. 1969. Organization of retina of the mudpuppy, *Necturus maculosus*. I. Synaptic structure. *J. Neurophysiol.* 32:315–338.

Dubin, M. 1970. The inner plexiform layer of the vertebrate retina: a quantitative and comparative electron microscopic analysis. *J. Comp. Neurol.* 140:479–506.

Duke-Elder, S. 1963. *System of Ophthalmology: Normal and Abnormal Development*, vol. 3. London: Klimpton.

Eccles, J. C. 1957. *The Physiology of Nerve Cells*. Baltimore: Johns Hopkins University Press.

———— 1964. *The Physiology of Synapses.* New York: Academic Press.

———— 1975. Under the spell of the synapse. In *The Neurosciences: Paths of Discovery*, ed. F. G. Worden, J. P. Swazey, and G. Adelman. Cambridge, Mass.: MIT Press.

Ehinger, B. 1966. Adrenergic retinal neurons. *Z. Zellforsch. Mikrosk. Anat.* 71:146–152.

———— 1976. Biogenic monoamines as transmitters in the retina. In *Transmitters in Visual Process*, ed. S. L. Bonting. Oxford: Pergamon Press, pp. 145–163.

———— 1982. Neurotransmitter systems in the retina. *Retina* 2:305–321.

Ehinger, B., B. Falck, and A. M. Laties. 1969. Adrenergic neurons in teleost retina. *Z. Zellforsch. Mikrosk. Anat.* 97:285–297.

Ehinger, B., and I. Holmgren. 1979. Electron microscopy of the indoleamine-accumulating neurons in the retina of the rabbit. *Cell Tissue Res.* 197:175–194.

Elliott, T. R. 1904. On the action of adrenalin. *J. Physiol.* 31:xx (Proc.).

Emeis, D., H. Kuhn, J. Reichert, and K. P. Hofmann. 1982. Complex formation between metarhodopsin II and GTP-binding protein in bovine photoreceptor membranes leads to a shift of the photoproduct equilibrium. *FEBS Lett.* 143:29–34.

Enroth-Cugell, C., and J. G. Robson. 1966. The contrast sensitivity of retinal ganglion cells of the cat. *J. Physiol.* 187:517–552.

Faber, D. S. 1969. Analysis of the slow transretinal potentials in response to light. Ph.D. dissertation. University of New York, Buffalo.

Fain, G. L. 1975a. Interactions of rod and cone signals in the mudpuppy retina. *J. Physiol.* 252:735–769.

———— 1975b. Quantum sensitivity of rods in the toad retina. *Science* 187:838–841.

———— 1976. Sensitivity of toad rods: dependence on wave-length and background illumination. *J. Physiol.* 261:71–101.

———— 1977. The threshold signal of photoreceptors. In *Vertebrate Photoreception*, ed. H. B. Barlow and P. Fatt. London: Academic Press, pp. 305–323.

Fain, G., and Dowling, J. E. 1973. Intracellular recordings from single rods and cones in the mudpuppy retina. *Science* 180:1178–1181.

Fain, G. L., G. H. Gold, and J. E. Dowling. 1976. Receptor coupling in the toad retina. *Cold Spring Harbor Symp. Quant. Biol.* 40:547–561.

Fain, G. L., A. M. Granda, and J. H. Maxwell. 1977. The voltage signal of photoreceptors at the visual threshold. *Nature* 265:181–183.

Fain, G. L., F. N. Quandt, B. L. Bastian, and H. M. Gerschenfeld. 1978. Contribution of a caesium-sensitive conductance increase to the rod response. *Nature* 272:467–469.

Falk, G., and P. Fatt. 1972. Physical changes induced by light in the rod outer segment of vertebrates. In *Handbook of Sensory Physiology: Photochemistry of Vision*, vol. 7, pt. 1, ed. H. J. A. Dartnell. New York: Springer-Verlag, pp. 200–244.

Famiglietti, E. V., Jr. 1983. On and off pathways through amacrine cells in mammalian retina: the synaptic connections of "starburst" amacrine cells. *Vision Res.* 23:1265–1279.

Famiglietti, E. V., Jr., A. Kaneko, and M. Tachibana. 1977. Neuronal architecture of on and off pathways to ganglion cells in carp retina. *Science* 198:1267–1269.

Famiglietti, E. V., Jr., and H. Kolb. 1975. A bistratified amacrine cell and synaptic circuitry in the inner plexiform layer of the retina. *Brain Res.* 84:293–300.

———— 1976. Structural basis for on- and off-center responses in retinal ganglion cells. *Science* 194:193–195.

Fesenko, E. E., S. S. Kolenikov, and A. L. Lyuborsky. 1985. Induction by cyclic GMP of cationic conductance in plasma membrane of retinal rod outer segment. *Nature* 313:310–313.

Fisher, S. K., and B. B. Boycott. 1974. Synaptic connexions made by horizontal cells within the outer plexiform layer of the retina of the cat and the rabbit. *Proc. R. Soc. Lond. B* 186:317–331.

Flamm, R. E., and R. M. Harris-Warrick. 1986a. Aminergic modulation in lobster stomatogastric ganglion. I. Effects on motor pattern and activity of neurons within the pyloric circuit. *J. Neurophysiol.* 55:847–865.

———— 1986b. Aminergic modulation in lobster stomatogastric ganglion. II. Target neurons of dopamine, octopamine, and serotonin within the pyloric circuit.

J. Neurophysiol. 55:866–881.

Flock, Å. 1964. Structure of the *Macula utriculi* with special reference to directional interplay of sensory responses as revealed by morphological polarization. *J. Cell Biol.* 22:413–431.

Frank, R. N., and J. E. Dowling. 1968. Rhodopsin photoproducts: effects on electroretinogram sensitivity in isolated perfused rat retina. *Science* 161:487–489.

Frederick, J. M., M. E. Rayborn, A. M. Laties, D. M.-K. Lam, and J. G. Hollyfield. 1982. Dopaminergic neurons in the human retina. *J. Comp. Neurol.* 210:65–79.

Frumkes, T. E., R. F. Miller, M. Slaughter, and R. F. Dacheux. 1981. Physiological and pharmacological basis of GABA and glycine action on neurons of mudpuppy retina. III. Amacrine-mediated inhibitory influences on ganglion cell receptive-field organization: a model. *J. Neurophysiol.* 45:783–804.

Fulton, A. B., and W. A. H. Rushton. 1978. The human rod ERG: correlation with psychophysical responses in light and dark adaptation. *Vision Res.* 18:793–800.

Fung, B. K.-K., J. B. Hurley, and L. Stryer. 1981. Flow of information in the light-triggered cyclic nucleotide cascade of vision. *Proc. Natl. Acad. Sci. USA* 78:152–156.

Fung, B., and L. Stryer. 1980. Photolyzed rhodopsin catalyzes the exchange of GTP for bound GDP in retinal rod outer segments. *Proc. Natl. Acad. Sci. USA* 77:2500–2504.

Fuortes, M. G. F., and E. J. Simon. 1974. Interactions leading to horizontal cell responses in the turtle retina. *J. Physiol.* 240:177–198.

Furshpan, E. J., and D. D. Potter. 1959. Transmission at the giant motor synapses of the crayfish. *J. Physiol.* 145:289–325.

Gallego, A. 1971. Horizontal and amacrine cells in the mammal's retina. *Vision Res.* 3:33–50.

——— 1982. Horizontal cells of the tetrapoda retina. In *The S-Potential,* ed. B. D. Drujan and M. Laufer. New York: Alan R. Liss, pp. 9–29.

Gerschenfeld, H. M., and M. Piccolino. 1979. Pharmacology of the connections of cones and L-horizontal cells in the vertebrate retina. In *The Neurosciences: Fourth Study Program,* ed. F. O. Schmitt and F. G. Worden.

Cambridge, Mass.: MIT Press, pp. 213–226.

Glickman, R. D., A. R. Adolph, and J. E. Dowling. 1982. Inner plexiform circuits in the carp retina: effects of cholinergic agonists, GABA, and substance on the ganglion cells. *Brain Res.* 234:81–99.

Gold, G. H. 1979. Photoreceptor coupling in retina of the toad, *Bufo marinus.* II. Physiology. *J. Neurophysiol.* 42:311–328.

——— 1986. Plasma membrane calcium fluxes are inconsistent with the Ca-hypothesis. *Proc. Natl. Acad. Sci. USA* 83:1150–1154.

Gold, G. H., and J. E. Dowling. 1979. Photoreceptor coupling in retina of the toad, *Bufo marinus.* I. Anatomy. *J. Neurophysiol.* 42:292–310.

Goldstein, E. B., and E. L. Berson. 1969. Cone dominance of the human early receptor potential. *Nature* 222:1272–1273.

Gouras, P. 1968. Identification of cone mechanisms in monkey ganglion cells. *J. Physiol.* 199:533–547.

Granit, R. 1933. The components of the retinal action potentials and their relation to the discharge in the optic nerve. *J. Physiol.* 77:207–240.

——— 1947. *Sensory Mechanisms of Retina.* London: Oxford University Press.

——— 1955. *Receptors and Sensory Perception.* New Haven: Yale University Press.

Granit, R., A. Munsterhjelm, and M. Zewi. 1939. The relation between concentration of visual purple and retinal sensitivity to light during dark adaptation. *J. Physiol.* 96:31–44.

Gray, E. G. 1959. Axo-somatic and axo-dendritic synapses of the cerebral cortex: an electron microscope study. *J. Anat.* 93:420–433.

Gray, E. G., and H. L. Pease. 1971. On understanding the organization of the retinal receptor synapses. *Brain Res.* 35:1–15.

Green, D. G. 1973. Scotopic and photopic components of the rat electroretinogram. *J. Physiol.* 228:781–797.

Green, D. G., and J. E. Dowling. 1975. Electrophysiological evidence for rod-like receptors in the gray squirrel, ground squirrel and prairie dog retinas. *J. Comp. Neurol.* 159:461–472.

Green, D. G., J. E. Dowling, I. M. Siegal, and H. Ripps. 1975. Retinal mechanisms of visual adaptation in the skate. *J. Gen. Physiol.* 65:483–502.

Green, D. G., and M. K. Powers. 1982. Mechanisms of light adaptation in rat retina. *Vision Res.* 22:209–216.

Greengard, P. 1978. *Cyclic Nucleotides, Phosphorylated Proteins and Neuronal Function*. New York: Raven Press.

Häggendal, J., and T. Malmfors. 1963. Evidence of dopamine-containing neurons in the retina of rabbits. *Acta Physiol. Scand.* 59:295–296.

Hagins, W. A. 1979. Excitation in vertebrate photoreceptors. In *The Neurosciences: Fourth Study Program*, ed. F. O. Schmitt and F. G. Worden. Cambridge, Mass.: MIT Press, pp. 183–191.

Hagins, W. A., R. D. Penn, and S. Yoshikami. 1970. Dark current and photocurrent in retinal rods. *Biophys. J.* 10:380–412.

Harris, G. G., L. Frishkopf, and Å. Flock. 1970. Receptor potentials from hair cells of the lateral line. *Science* 167:76–79.

Hartline, H. K. 1938. The response of single optic nerve fibers of the vertebrate eye to illumination of the retina. *Am. J. Physiol.* 121:400–415.

Hartline, H. K., and F. Ratliff. 1957. Inhibitory interaction of receptor units in the eye of *Limulus*. *J. Gen. Physiol.* 40:357–376.

Hashimoto, Y., M. Abe, and M. Inokuchi. 1980. Identification of the interplexiform cell in the dace retina by dye-injection method. *Brain Res.* 197:331–340.

Hassin, G. 1979. Pikeperch horizontal cells identified by intracellular staining. *J. Comp. Neurol.* 186:529–540.

Haynes, L. W., and K. -W. Yau. 1985. Cyclic GMP-sensitive conductance in outer segment membranes of catfish cones. *Nature* 317:61–64.

Hecht, S. 1942. The chemistry of visual substances. *Annu. Rev. Biochem.* 11:465–496.

Hecht, S., C. Haig, and A. M. Chase. 1937. The influence of light-adaptation on subsequent dark-adaptation of the eye. *J. Gen. Physiol.* 20:831–850.

Hecht, S., S. Shlaer, and M. H. Pirenne. 1942. Energy, quanta, and vision. *J. Gen. Physiol.* 25:819–840.

Hedden, W. L., and J. E. Dowling. 1978. The interplexiform cell system. II. Effects of dopamine on goldfish retinal neurons. *Proc. R. Soc. Lond. B.* 201:27–55.

Holmgren, F. 1865–1866. Metod att objektivisera effekten av ljusintryck på retina. *Upsala Läkarefören. Förh.* 1:177–191.

Holmgren-Taylor, I. 1982. Electron microscopical observations on the indoleamine-accumulating neurons and their synaptic connections in the retina of the cat. *J. Comp. Neurol.* 208:144–156.

Honrubia, F. M., and J. H. Elliott. 1968. Efferent innervation of the retina. I. Morphological study of the human retina. *Arch. Ophthalmol.* 80:98–103.

Hosokawa, Y., and K.-I. Naka. 1985. Spontaneous membrane fluctuation in catfish type-N cells. *Vision Res.* 25:539–542.

Hubbard, R. 1953–54. The molecular weight of rhodopsin and the nature of the rhodopsin-digitonin complex. *J. Gen. Physiol.* 37:381–399.

Hubbard, R., and A. Kropf. 1958. The action of light on rhodopsin. *Proc. Natl. Acad. Sci. USA* 44:130–139.

Hubbard, R., and G. Wald. 1952–1953. Cis–trans isomers of vitamin A and retinene in rhodopsin synthesis. *J. Gen. Physiol.* 36:269–315.

Hubel, D. H., and T. N. Wiesel. 1959. Receptive fields of single neurons in the cat's striate cortex. *J. Physiol.* 148:574–591.

—— 1960. Receptive fields of optic nerve fibers in the spider monkey. *J. Physiol.* 154:572–580.

——1962. Receptive fields, binocular interaction and functional architecture in the cat's visual cortex. *J. Physiol.* 160:106–154.

—— 1968. Receptive fields and functional architecture of monkey striate cortex. *J. Physiol.* 195:215–243.

Ikeda, H., and M. J. Sheardown. 1982. Aspartate may be an excitatory transmitter mediating visual excitation of "sustained" but not "transient" cells in the cat retina: iontophoretic studies *in vivo*. *Neuroscience* 7:25–36.

Ishida, A. T., and G. L. Fain. 1981. D-Aspartate potentiates the effects of L-glutamate on horizontal cells in goldfish retina. *Proc. Natl. Acad. Sci. USA* 78:5890–5894.

Ishida, A. T., A. Kaneko, and M. Tachibana. 1984. Responses of solitary retinal horizontal cells from *Carassius auratus* to L-glutamate and related amino acids. *J. Physiol.* 348:255–270.

Ishida, A. T., W. K. Stell, and D. O. Lightfoot. 1980. Rod and cone inputs to bipolar cells in goldfish retina. *J. Comp. Neurol.* 191:315–335.

Itzhaki, A., and I. Perlman. 1987. Light adaptation of red cones and L1-horizontal cells in the turtle retina: effect of the background spatial pattern. *Vision Res.,* in press.

Jan, L. Y., and Y. N. Jan. 1982. Peptidergic transmission in sympathetic ganglia of the frog. *J. Physiol.* 327: 219–246.

Jan, L., and J. P. Revel. 1973. Localization of rhodopsin in the retina by electron microscopy. *J. Cell Biol.* 59:155.

Jensen, R. J., and N. W. Daw. 1984. Effects of dopamine antagonists on receptive fields of brisk cells and directionally selective cells in the rabbit retina. *J. Neurosci.* 4:2972–2985.

Kandel, E. R. 1976. *Cellular Basis of Behavior.* San Francisco: W. H. Freeman.

Kandel, E. R., and J. H. Schwartz. 1982. Molecular biology of learning: modulation of transmitter release. *Science* 218:433–443.

Kaneko, A. 1970. Physiological and morphological identification of horizontal, bipolar, and amacrine cells in the goldfish retina. *J. Physiol.* 207:623–633.

——— 1971a. Electrical connexions between horizontal cells in the dogfish retina. *J. Physiol.* 213:95–105.

———1971b. Physiological studies of single retinal cells and their morphological identification. *Vision Res. Suppl.* 3:17–26.

——— 1973. Receptive field organization of bipolar and amacrine cells in the goldfish retina. *J. Physiol.* 235:133–153.

Kaneko, A., and H. Hashimoto. 1967. Recording site of the single cone response determined by electrode marking technique. *Vision Res.* 1:847–851.

Kaneko, A., and A. E. Stuart. 1984. Coupling between horizontal cells in the carp retina revealed by diffusion of Lucifer Yellow. *Neurosci. Lett.* 47:1–7.

Kaneko, A. and M. Tachibana. 1983. Double color-opponent receptive fields of carp bipolar cells. *Vision Res.* 23:381–388.

——— 1985a. A voltage-clamp analysis of membrane currents in solitary bipolar cells dissociated from *Caras-*

sius auratus. J. Physiol. 358:131–152.

——— 1985b. Effects of L-glutamate on the anomalous rectifier potassium current in horizontal cells of *Carassius auratus* retina. *J. Physiol.* 358:169–182.

Karten, H. J., and N. Brecha. 1980. Localization of substance P immunoreactivity in amacrine cells of the retina. *Nature* 283:87–88.

——— 1982. Neuropeptides in the vertebrate retina. In *Neurotransmitter Interaction and Compartmentation,* ed. H. F. Bradford. New York: Plenum, pp. 719–733.

Karwoski, C. J., and L. M. Proenza. 1977. Relationship between Müller cell responses, a local transretinal potential and potassium flux. *J. Neurophysiol.* 40:244–259.

——— 1978. Light-evoked changes in extracellular potassium concentration in mudpuppy retina. *Brain Res.* 142:515–530.

Katz, B., and R. Miledi. 1967. A study of synaptic transmission in the absence of nerve impulses. *J. Physiol.* 192:407–436.

Katz, B., and S. Thesleff. 1957. A study of the "desensitization" produced by acetylcholine at the motor endplate. *J. Physiol.* 138:63–80.

Kidd, M. 1962. Electron microscopy of the inner plexiform layer of the retina in the cat and the pigeon. *J. Anat.* 96:179–188.

Kirby, A. W., and C. Enroth-Cugell. 1976. The involvement of GABA in the organization of cat retinal ganglion cells. *J. Gen. Physiol.* 68:465–484.

Kleinschmidt, J. 1973. Adaptation properties of intracellularly recorded *Gekko* photoreceptor potentials. In *Biochemistry and Physiology of Visual Pigments,* ed. H. Langer. Berlin: Springer-Verlag, pp. 219–228.

Kleinschmidt, J., and J. E. Dowling. 1975. Intracellular recordings from gecko photoreceptors during light and dark adaptation. *J. Gen. Physiol.* 66:617–648.

Kleinschmidt, J., and S. Yazulla. 1984. Uptake of ^{3}H-glycine in the outer plexiform layer of the retina of the toad, *Bufo marinus. J. Comp. Neurol.* 230: 352–360.

Kline, R. P., H. Ripps, and J. E. Dowling. 1978. Generation of b-wave currents in the skate retina. *Proc. Natl. Acad. Sci. USA* 75:5727–5731.

———— 1985. Light-induced potassium fluxes in the skate retina. *Neuroscience* 14:225–235.

Knapp, A. G., and J. E. Dowling. 1987. Dopamine enhances excitatory amino acid-gated conductances in retinal horizontal cells. *Nature* 325:437–439.

Koch, C., T. Poggio, and V. Torre. 1982. Retinal ganglion cells: a functional interpretation of dendritic morphology. *Philos. Trans. R. Soc. Lond. B* 298:227–264.

Kolb, H. 1970. Organization of the outer plexiform layer of the primate retina: electron microscopy of Golgi-impregnated cells. *Philos. Trans. R. Soc. Lond. B* 258:261–283.

———— 1974. The connections between horizontal cells and photoreceptors in the retina of the cat: electron microscopy of Golgi preparations. *J. Comp. Neurol.* 155:1–14.

———— 1977. The organization of the outer plexiform layer in the retina of the cat: electron microscopic observations. *J. Neurocytol.* 6:131–153.

———— 1979. The inner plexiform layer in the retina of the cat: electron microscopic observations. *J. Neurocytol.* 8:295–329.

———— 1982. The morphology of the bipolar cells, amacrine cells and ganglion cells in the retina of the turtle *Pseudemys scripta elegans. Philos. Trans. R. Soc. Lond. B* 298:355–393.

Kolb, H., and J. Jones. 1982. Light and electron microscopy of the photoreceptors in the retina of the red-eared slider, *Pseudemys scripta elegans. J. Comp. Neurol.* 209:331–338.

———— 1984. Synaptic organization of the outer plexiform layer of the turtle retina: an electron microscope study of serial sections. *J. Neurocytol.* 13:567–591.

———— 1985. Electron microscopy of Golgi-impregnated photoreceptors reveals connections between red and green cones in the turtle retina. *J. Neurophysiol.* 54:304–317.

Kolb, H., R. Nelson, and A. Mariani. 1981. Amacrine cells, bipolar cells and ganglion cells of the cat retina: a Golgi study. *Vision Res.* 21:1081–1114.

Kolb, H., and R. West. 1977. Synaptic connections of the interplexiform cell in the retina of the cat. *J. Neurocytol.* 6:155–170.

Kraft, T. W., and D. A. Burkhardt. 1986. Telodendrites of cone photoreceptors: structure and probable function. *J. Comp. Neurol.* 249:13–27.

Kramer, S. G. 1971. Dopamine: a retinal neurotransmitter. I. Retinal uptake, storage and light-stimulated release of H3-dopamine in the retina. *Invest. Ophthalmol.* 10:617–624.

Kuffler, S. W. 1953. Discharge patterns and functional organization of mammalian retina. *J. Neurophysiol.* 16:37–68.

Kuffler, S. W., and J. G. Nicholls. 1966. The physiology of neuroglial cells. *Ergeb. Physiol. Exp. Pharmakol.* 57:1–90.

Kühn, H., N. Bennett, M. Michel-Villaz, and M. Chabre. 1981. Interactions between photoexcited rhodopsin and GTP-binding protein: kinetics and stoichiometric analysis from light scattering changes. *Proc. Natl. Acad. Sci. USA* 78:6873–6877.

Kühn, H., and W. J. Dreyer. 1972. Light-dependent phosphorylation of rhodopsin by ATP. *FEBS Lett.* 20:1–6.

Kühn, H., S. W. Hall, and U. Wilden. 1984. Light induced binding of 48-KDa protein to photoreceptor membranes is highly enhanced by phosphorylation of rhodopsin. *FEBS Lett.* 176:473–478.

Kühne, W. 1878. *On the Photochemistry of the Retina and on Visual Purple*, edited with notes by M. Foster. London: Macmillan.

Kujiraoka, T., and T. Saito. 1986. Electrical coupling between bipolar cells in the carp retina. *Proc. Natl. Acad. Sci. USA* 83:4063–4066.

Kusano, K. 1970. Influence of ionic environment on the relationship between pre- and postsynaptic potentials. *J. Neurobiol.* 1:435–457.

Ladman, A. J. 1958. The fine structure of the rod bipolar synapse in the retina of the albino rat. *J. Biophys. Biochem. Cytol.* 4:459–466.

Lam, D. M.-K. 1972. Biosynthesis of acetylcholine in turtle photoreceptors. *Proc. Natl. Acad. Sci. USA* 69:1987–1991.

———— 1975a. Biosynthesis of γ-aminobutyric acid by isolated axons of cone horizontal cells in the goldfish retina. *Nature* 254:345–347.

———— 1975b. Synaptic chemistry of identified cells in the vertebrate retina. *Cold Spring Harbor Symp. Quant. Biol.* 40:571–579.

Lam, D. M.-K., E. M. Laster, and K.-I. Naka. 1978. γ-Aminobutyric acid: a neurotransmitter candidate for cone horizontal cells of the catfish retina. *Proc. Natl. Acad. Sci. USA* 75:6310–6313.

Lam, D. M.-K., and L. Steinman. 1971. The uptake of [γ³H]-aminobutyric acid in the goldfish retina. *Proc. Natl. Acad. Sci. USA* 68:2777–2781.

Lamb, T. D., P. MacNaughton, and K.-W. Yau. 1981. Spatial spread of activation and background desensitization in toad rod outer segments. *J. Physiol.* 319:463–496.

Lamb, T. D., and E. J. Simon. 1976. The relation between intercellular coupling and electrical noise in turtle photoreceptors. *J. Physiol.* 263:257–286.

Lasansky, A. 1969. Basal junctions at synaptic endings of turtle visual cells. *J. Cell Biol.* 40:577–581.

——— 1971. Synaptic organization of the cone cells in the turtle retina. *Philos. Trans. R. Soc. Lond. B* 262:365–381.

——— 1973. Organization of the outer synaptic layer in the retina of the larval tiger salamander. *Philos. Trans. R. Soc. Lond. B* 265:471–489.

——— 1978. Contacts between receptors and electrophysiologically identified neurones in the retina of the larval tiger salamander. *J. Physiol.* 285:531–542.

Lasater, E. M. 1982. Spatial receptive fields of catfish retinal ganglion cells. *J. Neurophysiol.* 48:823–835.

——— 1986. Ionic currents of cultured horizontal cells isolated from white perch retina. *J. Neurophysiol.* 55:499–513.

Lasater, E. M., and J. E. Dowling. 1982. Carp horizontal cells in culture respond selectively to L-glutamate and its agonists. *Proc. Natl. Acad. Sci. USA* 79:936–940.

——— 1985a. Dopamine decreases conductance of the electrical junctions between cultured retinal horizontal cells. *Proc. Natl. Acad. Sci. USA* 82:3025–3029.

——— 1985b. Electrical coupling between pairs of isolated fish horizontal cells is modulated by dopamine and cyclic AMP. In *Gap Junctions*, ed. M. V. L. Bennett and D. C. Spray. Cold Spring Harbor, N.Y.: *Cold Spring Harbor Laboratory*, pp. 393–404.

Lasater, E. M., J. E. Dowling, and H. Ripps. 1984. Pharmacological properties of isolated horizontal and bipolar cells from the skate retina. *J. Neurosci.* 4:1966–1975.

Laties, A. M., and D. Jacobowitz. 1966. A comparative study of the autonomic innervation of the eye in monkey, cat and rabbit. *Anat. Rec.* 156:383–396.

Leeper, H. F., and D. R. Copenhagen. 1982. Horizontal cells in turtle retina: structure, synaptic connections, and visual processing. In *The S-Potential*, ed. B. D. Drujan and M. Laufer. New York: Alan R. Liss, pp. 77–104.

Levick, W. R. 1967. Receptive fields and trigger features of ganglion cells in the visual streak of the rabbit retina. *J. Physiol.* 188:285–307.

Li, H.-B., D. W. Marshak, J. E. Dowling, and D. M.-K. Lam. 1986. Colocalization of immunoreactive substance P and neurotensin in amacrine cells of the goldfish retina. *Brain Res.* 366:307–313.

Li, H.-B., C. B. Watt, and D. M.-K. Lam. 1985. The coexistence of two neuroactive peptides in a subpopulation of retinal amacrine cells. *Brain Res.* 345:176–180.

Liebman, P. A., and E. N. Pugh. 1980. ATP mediates rapid reversal of cyclic GMP phosphodiesterase activation in visual receptor membranes. *Nature* 287:734–736.

Lipton, S. A., H. Rasmussen, and J. E. Dowling. 1977. Electrical and adaptive properties of rod photoreceptors in *Bufo marinus*. II. Effects of cyclic nucleotides and prostaglandins. *J. Gen. Physiol.* 70:771–791.

Loewi, O. 1921. Über humorale Übertragbarkeit de Herznervenwirking. *Pflügers Arch.* 189:239–242.

MacNichol, E. F., and G. Svaetichin. 1958. Electric responses from the isolated retinas of fishes. *Am. J. Ophthalmol.* 46:26–46.

Maksimova, Y. M. 1970. Effects of intracellular polarization of horizontal cells on the activity of the ganglion cells of the retina of fish. *Biofizika* 14:570–577.

Mangel, S. C., M. Ariel, and J. E. Dowling. 1985. Effects of amino acid antagonists upon the spectral properties of carp horizontal cells: circuitry of the outer retina. *J. Neurosci.* 5:2839–2850.

Mangel, S. C., and J. E. Dowling. 1985. Responsiveness and receptive field size of carp horizontal cells are reduced by prolonged darkness and dopamine. *Science* 229:1107–1109.

Mann, I. C. 1964. *The Development of the Human Eye.* London: British Medical Association.

Marc, R. E., and D. M.-K. Lam. 1981. Uptake of aspartic

and glutamic acids by photoreceptors in the goldfish retina. *Proc. Natl. Acad. Sci. USA* 78:7185–7189.

Marc, R., and W. L. S. Liu. 1984. Horizontal cell synapses onto glycine-accumulating interplexiform cells. *Nature* 311:266–269.

Marc, R. E., W. K. Stell, D. Bok, and D. M.-K. Lam. 1978. GABA-ergic pathways in the goldfish retina. *J. Comp. Neurol.* 182:221–246.

Marchiafava, P. L. 1978. Horizontal cells influence membrane potential of bipolar cells in the retina of the turtle. *Nature* 275:141–142.

———1983. The organization of inputs establishes two functional and morphologically identifiable classes of ganglion cells in the retina of the turtle. *Vision Res.* 23:325–338.

Marchiafava, P. L., and R. Weiler. 1980. Intracellular analysis and structural correlates of the organization of inputs to ganglion cells in the retina of the turtle. *Proc. R. Soc. Lond. B* 208:103–113.

Mariani, A. P. 1981. A diffuse, invaginating cone bipolar cell in primate retina. *J. Comp. Neurol.* 197:661–671.

——— 1982. Biplexiform cells: ganglion cells of the primate retina that contact photoreceptors. *Science* 216:1134–1136.

Marks, W. B. 1965. Visual pigments of single goldfish cones. *J. Physiol.* 178:14–32.

Marks, W. B., W. H. Dobelle, and E. F. MacNichol. 1964. Visual pigments of single primate cones. *Science* 143:1181–1183.

Marshak, D. W., and J. E. Dowling. 1987. Synapses of cone horizontal cell axons in goldfish retina. *J. Comp. Neurol.* 256:430–443.

Masland, R. H., and A. Ames. III. 1976. Responses to acetylcholine of ganglion cells in an isolated mammalian retina. *J. Neurophysiol.* 39:1220–1235.

Masland, R. H., and C. J. Livingstone. 1976. Effect of stimulation with light on synthesis and release of acetylcholine by an isolated mammalian retina. *J. Neurophysiol.* 39:1210–1219.

Masland, R. H., and J. W. Mills. 1979. Autoradiographic identification of acetylcholine in the rabbit retina. *J. Cell Biol.* 83:159–178.

Masland, R. H., J. W. Mills, and C. Cassidy. 1984b. The functions of acetylcholine in the rabbit retina. *Proc. R. Soc. Lond. B.* 223:121–139.

Masland, R. H., J. W. Mills, and S. A. Hayden. 1984a. Acetylcholine-synthesizing amacrine cells: identification and selective staining using radioautography and fluorescent markers. *Proc. R. Soc. Lond. B* 223:79–100.

Massey, S. C., and M. J. Neal. 1979. The light evoked release of acetylcholine from rabbit retina *in vivo* and its inhibition by GABA. *J. Neurochem.* 32:1327–1329.

Massey, S. C., and D. A. Redburn. 1982. A tonic γ-aminobutyric acid-mediated inhibition of cholinergic amacrine cells in rabbit retina. *J. Neurosci.* 2:1633–1643.

Matsumoto, N., and K.-I. Naka. 1972. Identification of intracellular responses in the frog retina. *Brain Res.* 42:59–71.

Matsuura, T., W. H. Miller, and T. Tomita. 1978. Cone-specific c-wave in the turtle retina. *Vision Res.* 18:767–775.

Matthews, R. G., R. Hubbard, P. K. Brown, and G. Wald. 1963. Tautomeric forms of metarhodopsin. *J. Gen. Physiol.* 47:215–240.

Maturana, H. R., and S. Frenk. 1963. Directional movement and horizontal edge detectors in the pigeon retina. *Science* 142:977–979.

——— 1965. Synaptic connections of the centrifugal fibers in the pigeon retina. *Science* 150:359–361.

Maturana, H. R., J. Y. Lettvin, W. S. McCulloch, and W. H. Pitts. 1960. Anatomy and physiology of vision in the frog (*Rana pipiens*). *J. Gen. Physiol.* 43:129–175.

McGuire, B. A., J. K. Stevens, and P. Sterling. 1984. Microcircuitry of bipolar cells in cat retina. *J. Neurosci.* 4:2920–2938.

——— 1986. Microcircuitry of beta ganglion cells in cat retina. *J. Neurosci.* 6:907–918.

Michael, C. R. 1968a. Receptive fields of single optic nerve fibers in a mammal with an all-cone retina. I. Contrast-sensitive units. *J. Neurophysiol.* 31:249–256.

——— 1968b. Receptive fields of single optic nerve fibers in a mammal with an all-cone retina. II. Directionally selective units. *J. Neurophysiol.* 31:257–267.

——— 1968c. Receptive fields of single optic nerve fibers in a mammal with an all-cone retina. III. Opponent color units. *J. Neurophysiol.* 31:268–282.

Miki, N., J. J. Keirns, F. R. Marcus, J. Freeman, and M. W. Bitensky. 1973. Regulation of cyclic nucleotide concentrations in photoreceptors: an ATP-dependent stimulation of cyclic nucleotide phosphodiesterase by light. *Proc. Natl. Acad. Sci. USA* 70:3820–3824.

Miller, R. F. 1973. Role of K$^+$ in generation of b-wave of electroretinogram. *J. Neurophysiol.* 36:28–38.

——— 1979. The neuronal basis of ganglion cell receptive field organization and the physiology of amacrine cells. In *The Neurosciences, Fourth Study Program,* ed. F. O. Schmitt, and F. G. Worden. Cambridge, Mass.: MIT Press, pp. 227–245.

Miller, R. F., and R. F. Dacheux. 1973. Information processing in the retina: importance of chloride ions. *Science* 181:266–268.

——— 1976a. Dendritic and somatic spikes in mudpuppy amacrine cells: identification and TTX sensitivity. *Brain Res.* 104:157–162.

——— 1976b. Synaptic organization and ionic basis of on and off channels in mudpuppy retina. I. Intracellular analysis of chloride-sensitive electrogenic properties of receptors, horizontal cells, bipolar cells and amacrine cells. *J. Gen. Physiol.* 67:639–659.

——— 1976c. Synaptic organization and ionic basis of on and off channels in mudpuppy retina. II. Chloride-dependent ganglion cell mechanisms. *J. Gen. Physiol.* 67:661–678.

——— 1976d. Synaptic organization and ionic basis of on and off channels in mudpuppy retina. III. A model of ganglion cell receptive field organization based on chloride-free experiments. *J. Gen. Physiol.* 67:679–690.

——— 1983. Intracellular chloride in retinal neurons: measurement and meaning. *Vision Res.* 23:399–411.

Miller, R. F., and J. E. Dowling. 1970. Intracellular responses of the Müller (glial) cells of mudpuppy retina: their relation to b-wave of the electroretinogram. *J. Neurophysiol.* 33:323–341.

Miller, R. F., T. E. Frumkes, M. Slaughter, and R. F. Dacheux. 1981a. Physiological and pharmacological ba-

sis of GABA and glycine action on neurons of mudpuppy retina. I. Receptors, horizontal cells, bipolars, and G-cells. *J. Neurophysiol.* 45:743–763.

———1981b. Physiological and pharmacological basis of GABA and glycine action on neurons of mudpuppy retina. II. Amacrine and ganglion cells. *J. Neurophysiol.* 45:764–782.

Miller, W. H., and G. D. Nicol. 1979. Evidence that cyclic GMP regulates membrane potential of rod photoreceptors. *Nature* 280:64–66.

Missotten, L. 1965. *The Ultrastructure of the Retina.* Brussels: Arscia Uitgaven N.V..

Mitra, N. L. 1955. Quantitative analysis of cell types in mammalian neo-cortex. *J. Anat.* 89:467–483.

Muller, K. J., and U. J. McMahan. 1976. The shapes of sensory and motor neurones and the distribution of synapses in ganglia of the leech: a study using intracellular injection of horseradish peroxidase. *Proc. R. Soc. Lond. B* 194:481–499.

Murakami, M., and A. Kaneko. 1966. Differentiation of PIII subcomponents in cold-blooded vertebrate retinas. *Vision Res.* 6:627–636.

Murakami, M., K. Ohtsu, and T. Ohtsuka. 1972. Effects of chemicals on receptors and horizontal cells in the retina. *J. Physiol.* 227:899–913.

Murakami, M., T. Ohtsuka, and H. Shimazaki. 1975. Effects of aspartate and glutamate on the bipolar cells in the carp retina. *Vision Res.* 15:456–458.

Murakami, M., and Y. Shigematsu. 1970. Duality of conduction mechanism in bipolar cells of frog retina. *Vision Res.* 10:1–10.

Naka, K.-I. 1972. The horizontal cell. *Vision Res.* 12:573–588.

——— 1976. Neuronal circuitry in the catfish retina. *Invest. Ophthalmol.* 15:926–934.

——— 1977. Functional organization of catfish retina. *J. Neurophysiol.* 40:26–43.

——— 1980. A class of catfish amacrine cells responds preferentially to objects which move vertically. *Vision Res.* 20:961–965.

Naka, K.-I., R. Y. Chan, and S. Yasui. 1979. Adaptation in catfish retina. *J. Neurophysiol.* 42:441–454.

Naka, K.-I., and B. N. Christensen. 1981. Direct electrical connections between transient amacrine cells in the catfish retina. *Science* 214:462–464.

Naka, K.-I., and T. Ohtsuka. 1975. Morphological and functional identifications of catfish retinal neurons. II. Morphological identification. *J. Neurophysiol.* 38:72–91.

Naka, K.-I., and W. A. H. Rushton. 1966. S-potentials from colour units in the retina of fish (Cyprinidae). *J. Physiol.* 185:587–599.

———1967. The generation and spread of S-potentials in fish (Cyprinidae). *J. Physiol.* 192:437–461.

Naka, K.-I., and P. Witkovsky. 1972. Dogfish ganglion cell discharge resulting from extrinsic polarization of the horizontal cell. *J. Physiol.* 223:449–460.

Nakamura, Y., B. A. McGuire, and P. Sterling. 1980. Interplexiform cell in cat retina: identification by uptake of [³H]GABA and serial reconstruction. *Proc. Natl. Acad. Sci. USA* 77:658–661.

Nathans, J., T. P. Piantanida, R. L. Eddy, T. B. Shows, and D. S. Hogness. 1986b. Molecular genetics of inherited variation in human color vision. *Science* 232:203–210.

Nathans, J., D. Thomas, and D. S. Hogness. 1986a. Molecular genetics of human color vision: the genes encoding blue, green, and red pigments. *Science* 232:193–202.

Negishi, K., and B. Drujan. 1979. Reciprocal changes in center and surrounding S-potentials of fish retina in response to dopamine. *Neurochem. Res.* 4:313–318.

Negishi, K., S. Kato, T. Teranishi, and M. Laufer. 1978. An electrophysiological study on the cholinergic system in the carp retina. *Brain Res.* 148:85–93.

Negishi, K., T. Teranishi, and S. Kato. 1983. A GABA antagonist, bicuculline, exerts its uncoupling action on external horizontal cells through dopamine in carp retina. *Neurosci. Lett.* 37:261–266.

Nelson, R. 1973. A comparison of electrical properties of neurons in *Necturus* retina. *J. Neurophysiol.* 36:519–535.

——— 1977. Cat cones have rod input: a comparison of the response properties of cones and horizontal cell bodies in the retina of the cat. *J. Comp. Neurol.* 172:109–136.

——— 1982. AII amacrine cells quicken time course of rod signals in the cat retina. *J. Neurophysiol.* 47:928–947.

Nelson, R., E. V. Famiglietti, Jr., and H. Kolb. 1978. Intracellular staining reveals different levels of stratification for on- and off-center ganglion cells in cat retina. *J. Neurophysiol.* 41:472–483.

Nelson, R., and H. Kolb. 1983. Synaptic patterns and response properties of bipolar and ganglion cells in the cat retina. *Vision Res.* 23:1183–1195.

Nelson, R., H. Kolb, E. V. Famiglietti, Jr., and P. Gouras. 1976. Neural responses in the rod and cone systems of the cat retina: intracellular records and procion stains. *Invest. Ophthalmol.* 15:946–953.

Nelson, R., A. V. Lutzow, H. Kolb, and P. Gouras. 1975. Horizontal cells in cat retina with independent dendritic systems. *Science* 189:137–139.

Newman, E. A. 1980. Current source-density analysis of the b-wave of frog retina. *J. Neurophysiol.* 43:1355–1366.

——— 1984. Regional specialization of retinal glial cell membrane. *Nature* 309:155–157.

——— 1986. The Muller cell. In *Astrocytes,* vol. 1, ed. S. Federoff and A. Vernadakis. New York: Academic Press, pp. 149–171.

Newman, E. A., and L. L. Odette. 1984. Model of electroretinogram b-wave generation: a test of the K⁺ hypothesis. *J. Neurophysiol.* 51:164–182.

Nicholls, J. G., and D. A. Baylor. 1968. Specific modalities and receptive fields of sensory neurons in the CNS of the leech. *J. Neurophysiol.* 31:740–756.

Noell, W. K. 1953. Studies on the electrophysiology and the metabolism of the retina. Report number 1. USAF School of Aviation Medicine, Randolph Field, Texas.

——— 1954. The origin of the electroretinogram. *Am. J. Ophthalmol.* 28:78–90.

Normann, R. A., and I. Perlman. 1979a. The effects of background illumination on the photoresponses of red and green cones. *J. Physiol.* 286:491–507.

——— 1979b. Signal transmission from red cones to horizontal cells in the turtle retina. *J. Physiol.* 286:509–524.

Normann, R. A., I. Perlman, H. Kolb, J. Jones, and S. J. Doly. 1984. Direct excitatory interactions between cones of different spectral types in the turtle retina. *Science* 224:625–627.

Normann, R. A., and F. S. Werblin. 1974. Control of retinal sensitivity. I. Light and dark adaptation of vertebrate rods and cones. *J. Gen. Physiol.* 63:37–61.

Norton, A. L., H. Spekreijse, H. G. Wagner, and M. L. Wolbarsht. 1970. Responses to directional stimuli in retinal preganglionic units. *J. Physiol.* 206:93–107.

Nye, P. W. 1968. An examination of the electroretinogram of the pigeon in response to stimuli of different intensity and wavelength and following intense chromatic adaptation. *Vision Res.* 8:679–696.

Oakley, B., II, D. G. Flaming, and K. T. Brown. 1979. Effects of the rod receptor potential upon retinal extracellular potassium concentration. *J. Gen. Physiol.* 74:713–737.

Oakley, B., II, and D. G. Green. 1976. Correlation of light-induced charges in retinal extracellular potassium concentration with the c-wave of the electroretinogram. *J. Neurophysiol.* 39:1117–1133.

Oakley, B., II, R. H. Steinberg, S. S. Miller, and S. E. Nilsson. 1977. The *in vitro* frog pigment epithelial cell hyperpolarization in responses to light. *Invest. Ophthalmol. Vis. Sci.* 16:771–774.

O'Connor, P., S. J. Dorison, K. J. Watling, and J. E. Dowling. 1986. Factors affecting release of ³H-dopamine from perfused carp retina. *J. Neurosci.* 6:1857–1865.

Ogden, T. 1973. The oscillatory waves of the primate electroretinogram. *Vision Res.* 13:1059–1074.

———— 1978. Astrocytes in neurofibrillar layer. Golgi studies. *Invest. Ophthalmol. Vis. Sci.* 17:499–510.

Ohtsuka, T. 1978. Combination of oil droplets with different types of photoreceptor in a freshwater turtle, *Geoclemys reevesii. Sens. Proc.* 2:321–325.

———— 1985. Spectral sensitivities of seven morphological types of photoreceptors in the retina of the turtle, *Geoclemys reevesii. J. Comp. Neurol.* 237:145–154.

Oyster, C. W., and E. S. Takahashi. 1977. Interplexiform cells in rabbit retina. *Proc. R. Soc. Lond. B* 197:477–484.

Oyster, C. W., E. S. Takahashi, M. Cilluffo, and N. C. Brecha. 1986. Morphology and distribution of tyrosine hydroxylase-like immunoreactive neurons in the cat retina. *Proc. Natl. Acad. Sci. USA* 82:6335–6339.

Pak, W. L. 1965. Some properties of the early electrical response in the vertebrate retina. *Cold Spring Harbor Symp. Quant. Biol.* 30:493–499.

Palay, S. L. 1956. Synapses in the central nervous system. *J. Biophys. Biochem. Cytol.* 2:193–202.

Pappas, G. D., and S. G. Waxman. 1972. Synaptic fine structure–morphological correlates of chemical and electrotonic transmission. In *Structure and Function of Synapses,* ed. G. D. Pappas and D. P. Purpura. New York: Raven Press, pp. 1–43.

Parthe, V. 1982. Horizontal cells in the teleost retina. In *The S-Potential,* ed. B. D. Drujan and M. Laufer. New York: Alan R. Liss, pp. 31–49.

Penn, R. D., and W. A. Hagins. 1969. Signal transmission along retinal rods and the origin of the electroretinographic a-wave. *Nature* 223:201–205.

Pepperberg, D. R., P. K. Brown, M. Lurie, and J. E. Dowling. 1978. Visual pigment and photoreceptor sensitivity in the isolated skate retina. *J. Gen. Physiol.* 71:369–396.

Piccolino, M., J. Neyton, and H. M. Gerschenfeld. 1984. Decrease of gap junction permeability induced by dopamine and cyclic adenosine 3′,5′-monophosphate in horizontal cells of turtle retina. *J. Neurosci.* 4:2477–2488.

Piccolino, M., J. Neyton, P. Witkovsky, and H. M. Gerschenfeld. 1982. γ-Aminobutyric acid antagonists decrease junctional communication between L-horizontal cells of the retina. *Proc. Natl. Acad. Sci. USA* 79:3671–3675.

Pinching, A. J., and T. P. S. Powell. 1971. The neuropil of the glomeruli of the olfactory bulb. *J. Cell Sci.* 9:347–377.

Pirenne, M. H. 1962. Absolute thresholds and quantum effects. In *The Eye,* vol. II, ed. H. Davson. New York: Academic Press, pp. 123–140.

Polyak, S. L. 1941. *The Retina.* Chicago: University of Chicago Press.

Poo, M., and R. Cone. 1974. Lateral diffusion of rhodopsin in the photoreceptor membrane. *Nature* 247:438–441.

Pourcho, R. G. 1980. Uptake of [³H]glycine and [³H]GABA by amacrine cells in the cat retina. *Brain Res.* 198:333–346.

Proenza, L. M., and D. A. Burkhardt. 1973. Proximal negative response and retinal sensitivity in the mudpuppy, *Necturus maculosus. J. Neurophysiol.* 36:502–518.

Rakič, P. 1975. Local circuit neurons. *Neurosci. Res. Prog. Bull.* 13:291–446.

Rall, W., and G. M. Shepherd. 1968. Theoretical reconstruction of field potentials and dendrodendritic synaptic interactions in olfactory bulb. *J. Neurophysiol.* 31:884–915.

Rall, W., G. M. Shepherd, T. S. Reese, and M. W. Brightman. 1966. Dendrodendritic synaptic pathway for inhibition in the olfactory bulb. *Exp. Neurol.* 14:44–56.

Ralston, H. J., III. 1971. Evidence for presynaptic dendrites and a proposal for their mechanism of action. *Nature* 230:585–587.

——— 1979. Neuronal circuitry of the ventrobasal thalamus: the role of presynaptic dendrites. In *The Neurosciences, Fourth Study Program,* ed. F. O. Schmitt and F. G. Worden. Cambridge, Mass.: MIT Press, pp. 373–379.

Ramón y Cajal, S. 1892. La rétine des vertébrés. *La céllule* 9:119–257. For English translations, see S. A. Thorpe and M. Glickstein, Springfield, Ill.: C. C. Thomas, 1972, and D. Maguire and R. W. Rodieck, in *The Vertebrate Retina,* San Francisco: W. H. Freeman, 1973, pp. 775–904.

——— 1911. Histologie du System Nerveux de l'Homme et des Vertébrés. Paris: A. Malsine.

——— 1964. *Recollections of My Life,* trans. E. Horne Craigie. Cambridge, Mass.: MIT Press.

Ratliff, F. 1965. *Mach Bands: Quantitative Studies on Neural Networks in the Retina.* San Francisco: Holden-Day.

——— 1972. Contour and Contrast. *Sci. Am.* 226:90–101.

Raviola, E., and N. B. Gilula. 1973. Gap junctions between photoreceptor cells in the vertebrate retina. *Proc. Natl. Acad. Sci. USA* 70:1677–1681.

——— 1975. Intramembrane organization of specialized contacts in the outer plexiform layer of the retina: a freeze-fracture study in monkeys and rabbits. *J. Cell Biol.* 65:192–222.

Raviola, G., and E. Raviola. 1966. Light and electron microscopic observations on the inner plexiform layer of the rabbit retina. *Am. J. Anat.* 120:403–426.

Rayborn, M. E., P. V. Sarthy, D. M.-K. Lam, and J. G. Hollyfield. 1981. The emergence, localization and maturation of neurotransmitter systems during development of the retina in *Xenopus laevis.* II. Glycine. *J. Comp. Neurol.* 195:585–593.

Ripps, H., L. Mehaffey, III, and I. M. Siegel. 1981. Rhodopsin kinetics in the cat retina. *J. Gen. Physiol.* 77:317–334.

Ripps, H., M. Shakib, and E. D. MacDonald. 1976. Peroxidase uptake by photoreceptor terminals of the skate retina. *J. Cell Biol.* 70:86–96.

Ripps, H., and P. Witkovsky. 1985. Neuron–glia interaction in the brain and retina. Prog. *Ret. Res.* 4:181–219.

Rodieck, R. W. 1973. *The Vertebrate Retina: Principles of Structure and Function.* San Francisco: W. H. Freeman.

Rowe, J. S., and K. H. Ruddock. 1982. Hyperpolarization of retinal horizontal cells by excitatory amino acid neurotransmitters. *Neurosci. Lett.* 30:251–256.

Rushton, W. A. H. 1961. Rhodopsin measurement and dark-adaptation in a subject deficient in cone vision. *J. Physiol.* 156:193–205.

——— 1963a. A cone pigment in the protanope. *J. Physiol.* 168:345–359.

——— 1963b. Increment threshold and dark adaptation. *J. Opt. Soc. Am.* 53:104–109.

——— 1965a. A foveal pigment in the deuteranope. *J. Physiol.* 176:38–45.

——— 1965b. Visual adaptation. *Proc. R. Soc. Lond. B* 162:20–46.

Rushton, W. A. H., and R. D. Cohen. 1954. Visual purple level and the course of dark adaptation. *Nature* 173:301–304.

Saito, H. A. 1983. Morphology of physiologically identified X-, Y-, and W-type retinal ganglion cells of the cat. *J. Comp. Neurol.* 221:279–288.

Saito, T., and A. Kaneko. 1983. Ionic mechanisms underlying the responses of off-center bipolar cells in the carp retina. I. Studies on responses evoked by light. *J. Gen. Physiol.* 81:589–601.

Saito, T., H. Kondo, and J. Toyoda. 1979. Ionic mechanisms of two types of on-center bipolar cells in the carp retina. I. The responses to central illumination. *J. Gen. Physiol.* 73:73–90.

Saito, T., T. Kujiraoka, and T. Yonaha. 1983. Connections between photoreceptors and horseradish peroxidase-injected bipolar cells in the carp retina. *Vision Res.* 23:353–362.

Sakai, H., and K.-I. Naka. 1983. Synaptic organization involving receptor, horizontal and on- and off-center bipolar cells in the catfish retina. *Vision Res.* 23:339–351.

——— 1985. Novel pathway connecting the outer and inner vertebrate retina. *Nature* 315:570–571.

——— 1986. Synaptic organization of the cone horizontal cells in the catfish retina. *J. Comp. Neurol.* 245:107–115.

Sakai, H., K.-I. Naka, and J. E. Dowling. 1986. Ganglion cell dendrites are presynaptic in catfish retina. *Nature* 319:495–497.

Sakuranaga, M., and K.-I. Naka. 1985a. Signal transmission in the catfish retina. I. Transmission in the outer retina. *J. Neurophysiol.* 53:373–389.

——— 1985b. Signal transmission in the catfish retina. II. Transmission to type-N cell. *J. Neurophysiol.* 53:390–410.

——— 1985c. Signal transmission in the catfish retina. III. Transmission to type-C cell. *J. Neurophysiol.* 53:411–428.

Schaeffer, S. F., E. Raviola, and J. E. Heuser. 1982. Membrane specializations in the outer plexiform layer of the turtle retina. *J. Comp. Neurol.* 204:253–276.

Schiller, P. H. 1984. The connections of the retinal on and off pathways to the lateral geniculate nucleus of the monkey. *Vision Res.* 24:923–932.

Schmidt, R., and R. Steinberg. 1971. Rod-dependent intracellular responses to light recorded from the pigment epithelium of the cat retina. *J. Physiol.* 217:71–91.

Scholes, J. H. 1975. Colour receptors and their synaptic connexions in the retina of a cyprinid fish. *Philos. Trans. R. Soc. Lond. B* 270:61–118.

Schwartz, E. A. 1974. Responses of bipolar cells in the retina of the turtle. *J. Physiol.* 236:211–224.

——— 1975. Cones excite rods in the retina of the turtle. *J. Physiol.* 246:639–651.

——— 1976. Electrical properties of the rod syncytium in the retina of the turtle. *J. Physiol.* 257:379–406.

——— 1982. Calcium-independent release of GABA from isolated horizontal cells of the toad retina. *J. Physiol.* 323:211–227.

Sheills, R. A., G. Falk, and S. Naghshineh. 1981. Action of glutamate and aspartate analogues on rod horizontal and bipolar cells. *Nature* 294:592–594.

Shingai, R., and B. N. Christensen. 1983. Sodium and calcium currents measured in isolated catfish horizontal cells under voltage clamp. *Neuroscience* 10:893–897.

Siegler, M. V. S., and M. Burrows. 1980. Non-spiking interneurones and local circuits. *Trends Neurosci.* 3:73–77.

Sjöstrand, F. S. 1953. Ultrastructure of the outer segments of rods and cones of the eye as revealed by the electron microscope. *Cell. Comp. Physiol.* 42:15–44.

——— 1958. Ultrastructure of retinal rod synapses of the guinea pig eye as revealed by three-dimensional reconstructions from sections. *J. Ultrastruct. Res.* 2:122–170.

Slaughter, M. M., and R. F. Miller. 1981. 2-Amino-4-phosphonobutyric acid: a new pharmacological tool for retina research. *Science* 211:182–185.

——— 1983a. An excitatory amino acid antagonist blocks cone input to sign-conserving second-order retinal neurons. *Science* 219:1230–1232.

——— 1983b. Bipolar cells in the mudpuppy retina use an excitatory amino acid neurotransmitter. *Nature* 303:537–538.

——— 1983c. The role of excitatory amino acid transmitters in the mudpuppy retina: an analysis with kainic acid N-methyl aspartate. *J. Neurosci.* 3:1701–1711.

——— 1985. Identification of a distinct synaptic glutamate receptor on horizontal cells in mudpuppy retina. *Nature* 314:96–97.

Sloper, J. J. 1971. Dendrodendritic synapses in the primate motor cortex. *Brain Res.* 34:186–192.

Smith, C. A., and F. S. Sjöstrand. 1961. A synaptic structure in the hair cells of the guinea pig cochlea. *J. Ultrastruct. Res.* 5:523–556.

Smith, R. G., M. A. Freed, and P. Sterling. 1986. Microcircuitry of the dark-adapted cat retina: functional architecture of the rod–cone network. *J. Neurosci.* 6:3505–3517.

Steinberg, R. H., R. Schmidt, and K. T. Brown. 1970. Intracellular responses to light from cat pigment epithelium: origin of the electroretinogram C-wave. *Nature* 227:728–730.

Steinberg, R. H., and I. Wood. 1974. Pigment epithelial cell ensheathment of cone outer segments in the retina of the domestic cat. *Proc. R. Soc. Lond. B* 187:461–478.

Stell, W. K. 1965. Correlation of retinal cytoarchitecture and ultrastructure in Golgi preparations. *Anat. Rec.* 153:389–397.

——— 1967. The structure and relationship of horizontal cells and photoreceptor–bipolar synaptic complexes in goldfish retina. *Am. J. Anat.* 121:401–424.

——— 1975. Horizontal cell axons and axon terminals in goldfish retina. *J. Comp. Neurol.* 159:503–519.

——— 1978. Inputs to bipolar cell dendrites in goldfish retina. *Sens. Proc.* 2:339–349.

Stell, W. K., and F. I. Harosi. 1976. Cone structure and visual pigment content in the retina of the goldfish. *Vision Res.* 16:647–657.

Stell, W. K., A. T. Ishida, and D. O. Lightfoot. 1977. Structural basis for on- and off-center responses in the retina of the goldfish. *Science* 198:1269–1271.

Stell, W. K., R. Kretz, and D. O. Lightfoot. 1982. Horizontal cell connectivity in goldfish. In *The S-Potential*, ed. B. D. Drujan and M. Laufer. New York: Alan R. Liss, pp. 51–75.

Stell, W. K., and D. O. Lightfoot. 1975. Color-specific interconnections of cones and horizontal cells in the retina of the goldfish. *J. Comp. Neurol.* 159:473–502.

Stell, W., D. Marshak, T. Yamada, N. Brecha, and H. J. Karten. 1980. Peptides are in the eye of the beholder. *Trends Neurosci.* 3:292–295.

Stevens, J. K., T. L. Davis, N. Friedman, and P. Sterling. 1980. A systematic approach to reconstructing microcircuitry by electron microscopy of serial sections. *Brain Res. Rev.* 2:265–293.

Stiles, W. S. 1949. Increment thresholds and the mechanisms of colour vision. *Doc. Ophthalmol.* 3:138–165.

Stone, J. 1983. *Parallel Processing in the Visual System*. New York: Plenum Press.

Stone, J., and M. Fabian. 1966. Specialized receptive fields of the cat's retina. *Science* 152:1277–1279.

Stone, J., and Y. Fukuda. 1974. Properties of cat retinal

ganglion cells: a comparison of W-cells with X- and Y-cells. *J. Neurophysiol.* 37:722–748.

Stone, J., and K.-P. Hoffman. 1972. Very slow-conducting ganglion cells in the cat's retina: a major, new functional type? *Brain Res.* 43:610–616.

Stone, S., and P. Witkovsky. 1984. The actions of γ-aminobutyric acid, glycine and their antagonists upon horizontal cells of the *Xenopus* retina. *J. Physiol.* 353:249–264.

Stryer, L. 1986. Cyclic GMP cascade of vision. *Annu. Rev. Neurosci.* 9:87–119.

Su, Y.-Y. T., C. B. Watt, and D. M.-K. Lam. 1985. Opioid pathways in an avian retina. I. The content, biosynthesis, and release of met^5-enkephalin. *J. Neurosci.* 5:851–856.

Svaetichin G. 1953. The cone action potential. *Acta Physiol. Scand.* 29:565–600.

Svaetichin, G., and E. F. MacNichol. 1958. Retinal mechanisms for chromatic and achromatic vision. *Ann. N. Y. Acad. Sci.* 74:385–404.

Tachibana, M. 1981. Membrane properties of solitary horizontal cells isolated from goldfish retina. *J. Physiol.* 321:141–166.

——— 1983. Ionic currents of solitary horizontal cells isolated from goldfish retina. *J. Physiol.* 345:329–351.

——— 1985. Permeability changes induced by L-glutamate in solitary retinal horizontal cells isolated from *Carassius auratus*. *J. Physiol.* 358:153–167.

Tachibana, M., and A. Kaneko. 1984. γ-Aminobutyric acid acts at axon terminals of turtle photoreceptors: difference in sensitivity among cell types. *Proc. Natl. Acad. Sci. USA* 81:7961–7964.

Takeuchi, A., and N. Takeuchi. 1962. Electrical changes in pre- and postsynaptic axons of the giant synapse of *Loligo*. *J. Gen. Physiol.* 45:1181–1193.

Tauchi, M., and R. H. Masland. 1984. The shape and arrangement of the cholinergic neurons in the rabbit retina. *Proc. R. Soc. Lond. B* 223:101–119.

Teranishi, T., K. Negishi, and S. Kato. 1983. Dopamine modulates S-potential amplitude and dye-coupling between external horizontal cells in carp retina. *Nature* 301:243–246.

——— 1984. Regulatory effect of dopamine on spatial properties of horizontal cells in carp retina. *J. Neurosci.* 4:1271–1280.

Thibos, L. N., and F. S. Werblin. 1978. The response properties of the steady antagonistic surround in the mudpuppy retina. *J. Physiol.* 278:79–99.

Tomita, T. 1963. Electrical activity in the vertebrate retina. *J. Opt. Soc. Am.* 53:49–57.

——— 1965. Electrophysiological study of the mechanisms subserving color coding in the fish retina. *Cold Spring Harbor Symp. Quant. Biol.* 30:559–566.

——— 1970. Electrical activity of vertebrate photoreceptors. *Q. Rev. Biophys.* 3:179–222.

Toyoda, J. 1973. Membrane resistance changes underlying the bipolar cell response in the carp retina. *Vision Res.* 13:283–294.

Toyoda, J., H. Hashimoto, and K. Ohtsu. 1973. Bipolar–amacrine transmission in the carp retina. *Vision Res.* 13:295–307.

Toyoda, J., and T. Kujiraoka. 1982. Analyses of bipolar cell responses elicited by polarization of horizontal cells. *J. Gen. Physiol.* 79:131–145.

Toyoda, J.-I., T. Kujiraoka, and M. Fujimoto. 1982. The opponent color process and interaction of horizontal cells. In *The S-Potential*, ed. B. D. Drujan and M. Laufer. New York: Alan R. Liss, pp. 151–160.

Toyoda, J.-I., and K. Tonosaki. 1978. Effect of polarization of horizontal cells on the on-center bipolar cell of the carp retina. *Nature* 276:399–400.

Trifonov, Y. A. 1968. Study of synaptic transmission between the photoreceptor and the horizontal cell using electrical stimulation of the retina. *Biofizika* 13:809–817.

Trifonov, Y. A., and A. L. Byzov. 1965. The response of the cells generating S-potential on the current passed through the eyecup of the turtle. *Biofizika* 10:673–680.

Uchizono, K. 1967. Synaptic organization of the Purkinje cells in the cerebellum of the cat. *Exp. Brain Res.* 4:97–113.

Van Buskirk, R., and J. E. Dowling. 1981. Isolated horizontal cells from carp retina demonstrate dopamine-dependent accumulation of cyclic AMP. *Proc. Natl. Acad. Sci. USA* 78:7825–7829.

Vaney, D. I. 1984. 'Coronate' amacrine cells in the rabbit retina have the 'starburst' dendritic morphology. *Proc. R. Soc. Lond. B* 220:501–508.

Vaney, D. I., L. Peichl, and B. B. Boycott. 1981. Matching populations of amacrine cells in the inner nuclear and ganglion cell layers of the rabbit retina. *J. Comp. Neurol.* 199:373–391.

Vaughn, J. E., E. V. Famiglietti, Jr., R. P. Barber, K. Saito, E. Roberts, and C. E. Ribak. 1981. GABAergic amacrine cells in rat retina: immumocytochemical identification and synaptic connectivity. *J. Comp. Neurol.* 197:113–127.

Victor, J. D., and R. M. Shapley. 1979. The nonlinear pathway of Y ganglion cells in the cat retina. *J. Gen. Physiol.* 74:671–689.

Voaden, M. J. 1976. γ-Aminobutyric acid and glycine as retinal neurotransmitters. In *Transmitters in the Visual Process*, ed. S. L. Bonting. Oxford: Pergamon Press, pp. 107–125.

Wachtmeister, L. 1972. On the oscillatory potentials of the human electroretinogram in light and dark adaptation. *Acta Ophthalmol. Suppl.* 116:1–32.

Wachtmeister, L., and J. E. Dowling. 1978. The oscillatory potentials of the mudpuppy retina. *Invest. Ophthalmol. Vis. Sci.* 17:1176–1188.

Wald, G. 1935. Carotenoids and the visual cycle. *J. Gen. Physiol.* 19:351–371.

——— 1955. The photoreceptor process in vision. *Am. J. Ophthalmol.* 40:18–41.

——— 1958. The significance of vertebrate metamorphosis. *Science* 128:1481–1490.

Wald, G., and P. K. Brown. 1950. The synthesis of rhodopsin from retinene. *Proc. Natl. Acad. Sci. USA* 36:84–92.

Walls, G. L. 1942. *The Vertebrate Eye and Its Adaptive Radiation*. New York: Hafner.

Waloga, G., and W. L. Pak. 1976. Horizontal cell potentials: dependence on external sodium ion concentration. *Science* 191:964.

Wässle, H., B. B. Boycott, and R. B. Illing. 1981b. Morphology and mosaic of on- and off-beta cells in the cat retina and some functional considerations. *Proc. R. Soc. Lond. B* 212:177–195.

Wässle, H., L. Peichl, and B. B. Boycott. 1981a. Morphology and topography of on- and off-alpha cells in the cat retina. *Proc. R. Soc. Lond. B* 212:157–175.

Watling, K. J., and J. E. Dowling. 1981. Dopamine mechanisms in the teleost retina. I. Dopamine-sensitive adenylate cyclase in homogenates of carp retina: effects of agonists, antagonists and ergots. *J. Neurochem.* 36:559–568.

——— 1983. Effects of vasoactive intestinal peptides on cyclic AMP accumulation in intact pieces and isolated horizontal cells of the teleost retina. *J. Neurochem.* 41:1205–1213.

Watling K. J., and L. L. Iversen. 1981. Comparison of the binding of [³H]spiperone and [³H]domperidone in homogenates of mammalian retina and caudate nucleus. *J. Neurochem.* 37:1130–1143.

Watt, C. B., H.-B. Li, and D. M.-K. Lam. 1985a. The presence of three neuroactive peptides in putative glycinergic amacrine cells of an avian retina. *Brain Res.* 348:187–191.

Watt, C. B., Y.-Y. T. Su, and D. M.-K. Lam. 1984. Interactions between enkephalin and GABA in avian retina. *Nature* 311:761–763.

——— 1985b. Opioid pathways in an avian retina. *J. Neurosci.* 5:857–865.

Weight, F. F. 1974. Synaptic potentials resulting from conductance decreases. In *Synaptic Transmission,* ed. M. V. L. Bennett. New York: Raven Press, pp. 141–152.

Weiler, R., and A. K. Ball. 1984. Co-localization of neurotensin-like immunoreactivity and H-glycine uptake in sustained amacrine cells of turtle retina. *Nature* 311:759–761.

Weinstein, G. W., R. R. Hobson, and J. E. Dowling. 1967. Light and dark adaptation in the isolated rat retina. *Nature* 215:134–138.

Werblin, F. S. 1970. Response of retinal cells to moving spots: intracellular recording in *Necturus maculosus. J. Neurophysiol.* 33:342–351.

——— 1972. Lateral interactions at inner plexiform layer of vertebrate retina: antagonistic responses to change. *Science* 175:1008–1010.

——— 1974. Control of retinal sensitivity. II. Lateral interactions at the outer plexiform layer. *J. Gen. Physiol.* 63:62–87.

——— 1975. Regenerative hyperpolarization in rods. *J. Physiol.* 244:53–81.

——— 1977a. Regenerative amacrine cell depolarization and formation of on–off ganglion cell responses. *J. Physiol.* 264:767–785.

——— 1977b. Synaptic interactions mediating bipolar re-

sponse in the retina of the tiger salamander. In *Vertebrate Photoreception,* ed. H. B. Barlow and P. Fatt. London: Academic Press, pp. 205–230.

Werblin, F. S., and D. R. Copenhagen. 1974. Control of retinal sensitivity. III. Lateral interactions at the inner plexiform layer. *J. Gen. Physiol.* 63:88–110.

Werblin, F. S., and J. E. Dowling. 1969. Organization of the retina of the mudpuppy, *Necturus maculosus.* II. Intracellular recording. *J. Neurophysiol.* 32:339–355.

West, R. W. 1976. Light and electron microscopy of the ground squirrel retina: functional considerations. *J. Comp. Neurol.* 168:355–377.

——— 1978. Bipolar and horizontal cells of the gray squirrel retina: Golgi morphology and receptor connections. *Vision Res.* 18:129–136.

West, R. W., and J. E. Dowling. 1972. Synapses onto different morphological types of retinal ganglion cells. *Science* 178:510–512.

——— 1975. Anatomical evidence for cone and rod-like receptors in the gray squirrel, ground squirrel and prairie dog retinas. *J. Comp. Neurol.* 159:439–459.

Witkovsky, P. 1980. Excitation and adaptation in the vertebrate retina. *Curr. Top. Eye Res.* 2:1–66.

Witkovsky, P., and J. E. Dowling. 1969. Synaptic relationships in the plexiform layers of carp retina. *Z. Zellforsch. Mikrosk. Anat.* 100:60–82.

Witkovsky, P., F. E. Dudek, and H. Ripps. 1975. Slow PIII component of the carp electroretinogram. *J. Gen. Physiol.* 65:119–134.

Witkovsky, P., J. Nelson, and H. Ripps. 1973. Action spectra and adaptation properties of carp photoreceptors. *J. Gen. Physiol.* 61:401–423.

Witkovsky, P., W. G. Owen, and M. Woodworth. 1983. Gap junctions among the perikarya, dendrites, and axon terminals of the luminosity-type horizontal cell of the turtle retina. *J. Comp. Neurol.* 216:359–368.

Witkovsky, P., M. Shakib, and H. Ripps. 1974. Interreceptoral junctions in the teleost retina. *Invest. Ophthalmol.* 13:996–1009.

Witkovsky, P., and W. K. Stell. 1973. Retinal structure in the smooth dogfish *Mustelus canis:* electron microscopy of serially sectioned bipolar cell synaptic terminals. *J. Comp. Neurol.* 150:147–168.

Woodruff, M. L., and M. D. Bownds. 1979. Amplitude,

kinetics and reversibility of a light-induced decrease in guanosine 3′,5′cyclic monophosphate in frog photoreceptor membranes. *J. Gen. Physiol.* 73:629–653.

Wyatt, H. J., and N. W. Daw. 1975. Directionally sensitive ganglion cells in the rabbit retina: specificity for stimulus direction, size, and speed. *J. Neurophysiol.* 38: 613–626.

Yamada, E., and T. Ishikawa. 1965. The fine structure of the horizontal cells in some vertebrate retinae. *Cold Spring Harbor Symp. Quant. Biol.* 30:383–392.

Yamada, T., D. Marshak, S. Basinger, J. Walsh, and J. Morley. 1980. Somatostatin-like immunoreactivity in the retina. *Proc. Natl. Acad. Sci. USA* 77:1691–1695.

Yau, K.-W., and K. Nakatani. 1985. Light-induced reduction of cytoplasmic free calcium in retinal rod outer segment. *Nature* 313:579–582.

Yazulla, S. 1976. Cone input to bipolar cells in the turtle retina. *Vision Res.* 16:737–744.

——— 1985. Evoked efflux of [³H]GABA from goldfish retina in the dark. *Brain Res.* 325:171–180.

Yazulla, S., and J. Kleinschmidt. 1982. Dopamine blocks carrier-mediated release of GABA from retinal horizontal cells. *Brain Res.* 233:211–215.

——— 1983. Carrier-mediated release of GABA from retinal horizontal cells. *Brain Res.* 263:63–75.

Yazulla, S., K. M. Studholme, and C. L. Zucker. 1985. Synaptic organization of substance P-like immunoreactive amacrine cells in goldfish retina. *J. Comp. Neurol.* 231:232–238.

Yonemura, D., Y. Masuda, and M. Hatta. 1963. The oscillatory potential in the electroretinogram. *Jpn. J. Physiol.* 13:129–137.

Index

Pages in italic refer to figures.